GENDER EQUITY SOURCES AND RESOURCES FOR EDUCATION STUDENTS

JO SANDERS
JANICE KOCH
JOSEPHINE URSO

Routledge
Taylor & Francis Group
LONDON AND NEW YORK

This book is based on work supported in part by the National Science Foundation under Grant No. HRD-9253182. Any opinions, findings, and conclusions or recommendations expressed in this publication are those of the authors and do not necessarily reflect the views of the National Science Foundation.

The work in this project was carried out at the Graduate School and University Center of the City University of New York.

First published by Lawrence Erlbaun Associates

2 Park Square, Milton Park, Abingdon, Oxon OX14 4RN
711 Third Avenue, New York, NY 10017, USA

Routledge is an imprint of the Taylor & Francis Group, an informa business

First issued in hardback 2017

Transferred to Digital Printing 2010

Library of Congress Cataloging-in-Publication Data

Sanders, Jo Shuchat.
Gender equity right from the start / Jo Sanders, Janice Koch, Josephine Urso.
p. cm.
Includes bibliographical references.
Contents: v. 1. Instructional activities for teacher educators in mathematics, science, and technology -- v. 2. Sources and resources for education students in mathematics, science, and technology.
ISBN 0-8058-2337-9 (v. 1 : alk. paper). -- ISBN 0-8058-2887-7 (v. 2 : alk. paper)
1. Educational equalization--United States. 2. Sex discrimination in education--United States. 3. Mathematics--Study and teaching--United States. 4. Science--Study and Teaching--United States. 5. Technology--Study and teaching--United States. 6. Textbook bias--United States. 7. Teachers--Training of--United States. 8. Women in mathematics--United States. 9. Women in science--United States. 10. Women in technology--United States. I. Koch, Janice, 1947- II. Urso, Josephine. III. Title.
LC213.2.S25 1997
379.2'6--dc21 97-21458
CIP

ISBN 978-0-8058-2887-0 (pbk)
ISBN 978-1-138-42877-5 (hbk)

CONTENTS

ACKNOWLEDGMENTS

Gender Equity Right From the Start is the beneficiary of many people's contributions.

We thank the funders of the Teacher Education Equity Project for their generosity and support: The National Science Foundation, IBM, Hewlett-Packard, and AT&T.

The 61 Teacher Education Equity Project participants tested, and in many cases improved on, the gender equity activities in *Gender Equity Right From the Start* with their preservice students. They have been wonderful to work with and are the heart of the project. Those who also contributed their own teaching activities to augment those of the authors are marked with an asterisk. They are:

Barbara Attivo, Wichita State University
Mike Beeth, Ohio State University
Huey Bogan, College of St. Rose
Robert Boram, Morehead State University
Rick Breault, University of Indianapolis
Cherry Brewton, Georgia Southern University
Patsy Brooks, Alverno College
Mike Cass, West Georgia College
Elisabeth Charron, Montana State University at Bozeman
Susan Chevalier, Adams State College
Frank Curriero, Jersey City State College
* **Clint Erb**, University of Vermont
Shirley Freed, Andrews University
Pamela Freeman, Mississippi State Univ
Donna Gee, Eastern New Mexico University
Maureen Gillette, College of St. Rose
* **Mike Grote**, Ohio Wesleyan University
* **Karen Higgins**, Oregon State University
S. Maxwell Hines, Hofstra University
Aurora Hodgden, Phillips University
Dave Johnson, Eastern Michigan University
* **Karen Karp**, University of Louisville
Gwen Kelly, University of Idaho
* **Jody Kenny**, St. Michael's College
Jerry Krockover, Purdue University
Shirley Leali, Univ. of Northern Colorado
* **Barb Levin**, University of North Carolina at Greensboro
Tom Lord, Indiana Univ. of Pennsylvania
* **Patricia Lucido**, Northwest Missouri State University
Sandy Madison, University of Wisconsin at Stevens Point
Kathi Matthew, Western Kentucky Univ.
Kathy Matthews, University of North Carolina at Greensboro
* **Leah McCoy**, Wake Forest University
Alice Mikovch, Western Kentucky Univ.
Rickie Miller, Boise State University
* **Joyce Morris**, University of Vermont
Rusty Myers, Alaska Pacific University
* **Maggie Niess**, Oregon State University
John Novak, Eastern Michigan University
George O'Brien, Florida International University
Ray Ostrander, Andrews University
Brenda Peters, College of St. Rose
Jenny Piazza, University of Southern Colorado
* **Charlie Rathbone**, University of Vermont
Joyce Saxon, Morehead State University
Dan Shepardson, Purdue University
Twyla Sherman, Wichita State University
Sheila Smith-Hobson, Lehman College
* **Yee-Ping Soon**, Florida International University
* **Dorothy Spethmann**, Dakota State Univ.
Rose Steelman, Univ. of Central Arkansas
Margaret Stempien, Indiana University of Pennsylvania
* **Korinne Tande**, Montana State University at Northern
Marie Theobald, University of Indianapolis

Meghan Twiest, Indiana University of Pennsylvania
* **Martha Voyles**, Grinnell College
Frank Walton, University of Hawaii
Ken Welty, University of Wisconsin at Stout
* **Judy Werner**, West Virginia University
* **Cathy Yeotis**, Wichita State University
Joe Zilliox, University of Hawaii

Several others also contributed activities, foremost among whom is Laura Jeffers who wrote a number of the technology activities, as well as Barbara Dannay and Marcy Ewell.

Larry Enochs and Marsha Lakes Matyas rescued the project from oblivion before it was funded when the National Science Foundation reorganized. We are immensely grateful for their tenaciousness.

The staff of the Teacher Education Equity Project have been incomparably talented and committed: Deirdre Armitage, Starla Rocco, and Dorothy Bozzone.

We are grateful to the City University of New York Graduate Center for providing a home for the project, and especially to Bert Flugman, the director of the Center for Advanced Study in Education, for his unfailing support.

The late Selma Greenberg was a pioneer in the area of gender equity in early childhood. We are indebted to her for the fine title from her book *Right From the Start: A Guide to Nonsexist Child Rearing*, published in 1978.

We are also grateful to the members of the Teacher Education Equity Project Advisory Committee:

Dennis Angle	Director, Kansas Careers, Kansas State University
Alice Artzt	Professor of Mathematics Education, Queens College
Henry Jay Becker	Professor of Education, University of California at Irvine
Marvin Druger	Past President, National Science Teachers Association, and Chair, Department of Science Teaching, Syracuse University
Elizabeth Fennema	Professor of Mathematics Education, University of Wisconsin at Madison
April Gardner	Assistant Professor of Biological Sciences, University of Northern Colorado
Peter Gerber	Director, Educational Programs, John D. & Catherine T. MacArthur Foundation
Lynn Glass	Past President, National Science Teachers Association, and Professor of Science Education, Iowa State University
Marilyn Guy	Past President, American Association of Colleges for Teacher Education, and Associate Professor of Education, Concordia College
David Imig	Chief Executive Officer, American Association of Colleges for Teacher Education

Jane Kahle	Condit Professor of Science Education, Miami University
Gavrielle Levine	Assistant Professor of Mathematics Education, Long Island University, C.W. Post Campus
Rebecca Lubetkin	Executive Director, Consortium for Educational Equity, Rutgers University
Bonnie Marks	Past President, International Society of Technology in Education, and Dirctor of Technology, Alameda County Office of Education, Hayward, California
Joann Jacullo Noto	Director of Teacher Education, Teachers College, Columbia University
Henry Olds	Senior Scientist, BBN Systems & Technologies, Cambridge, MA
Kate Scantlebury	Associate Professor of Chemistry and Biology, University of Delaware
Robert Sellar	Manager, Area External Programs, I.B.M. Corporation, New York
Charol Shakeshaft	Chair, Department of Administration and Policy Studies, School of Education, Hofstra University
Stan Silverman	Director, Educational Technology, New York Institute of Technology, Central Islip Campus
Lee Stiff	Board of Directors, National Council of Teachers of Mathematics, and Professor of Mathematics Education, North Carolina State University

Our deepest thanks are reserved for the evaluator of the Teacher Education Equity Project, Pat Campbell, for her absolute integrity, her sensitivity, her promptness and thoroughness, and her constant kindness.

Jo Sanders, Director, Gender Equity Program
Washington Research Institute, Seattle
Janice Koch, Hofstra University, Hempstead, Long Island
Josephine Urso, Community School District 15, Brooklyn, New York

INTRODUCTION

What We Know

Decades of research support the fact that classroom environments are experienced differently by males and females. As early as nursery school, boys and girls sitting in the same classroom, with the same teacher, using the same materials, have different learning experiences. These differences persist through their pre-college education and beyond. Frequently, these experiences marginalize the girls in the areas of mathematics, science, and technology.

What we know, to generalize broadly, is that in the early elementary years girls and boys do equally well in tests and grades in mathematics, science, and technology (MST). As females progress through school and into college and graduate school, despite their frequently higher course grades, they score lower on standardized tests than males and take fewer advanced courses, which means they drop out of mathematics, science, and/or technology earlier than males.

As a consequence, large numbers of women are not qualified to enter careers in science, mathematics, technology, and related fields. The underrepresentation is easily seen in the tables in the pages later in this section. It matters because on average, technical occupations yield considerably higher salaries *for the same amount of educational preparation*. Now that a single salary can rarely support an entire family and now that single-parent households are common, decent salaries are more important than ever. Technical occupations have relatively good career ladders. They are unusually varied: from technician to professional, and in academic, corporate, government, and nonprofit settings, as well as indoors and outdoors. And they are projected to grow well over the next decade: Computer engineers and scientists, for example, are expected to increase 112%. [1]

Education and employment trends concerning girls and women have been extensively studied and documented since the early 1970s. Considerable effort has been expended in reaching K-12 teachers since then with an awareness of the problem and knowledge of strategies that have been proven effective in increasing girls' participation in MST. In fact there has been measurable progress, although we are far from equality yet. But gender equity progress has not for the most part reached teacher education. New teachers thus enter classrooms every year unaware that there is a problem with girls and mathematics, science and technology, let alone how to address it. An exclusive emphasis on reaching inservice teachers makes no sense.

The reason it's important to reach preservice teachers is that gender bias in the classroom is nearly always inadvertent and, more often than you would think, below the level of consciousness. Research has shown, for example, that teachers call on boys more than girls in

[1] New York Times, September 3, 1995, page 9. Data from the Bureau of Labor Statistics, U.S. Dept. of Labor.

math, science and technology classes. Because teachers don't even realize they're doing it, they can't correct it. And although it really doesn't matter if boys are called on a little more than girls in one class period, the cumulative effect over years of schooling is to signal to children that boys' thoughts and answers are more valuable than those of girls. Indeed, it is the accumulation of subtle and unintended gender biases such as this that result in the severely lopsided occupational figures presented later in this section. It is not good for us as a society that women account for only 32% of chemists and 8% of engineers.

When girls and women fail to persist in mathematics, science and technology to the extent they otherwise could, this is not a women's problem. It is a human problem.

The Teacher Education Equity Project

Gender Equity Right From the Start was developed in the Teacher Education Equity Project, funded from 1993 to 1996 for $1,028,000 by the National Science Foundation, IBM, Hewlett-Packard, and AT&T. The project worked with 61 teacher educators in 40 colleges and universities in 27 states who teach methods courses in mathematics, science, and technology. Its goal was to help them teach their preservice students about gender equity. One third were men, and one sixth were people of color. They taught gender equity activities to their students, shared what they learned in the project with their colleagues, and carried out a mini-grant project. The 61 teacher educators tested most and wrote some of the activities in *Gender Equity Right From the Start.*

The project clearly accomplished what it set out to do. It had a stunning multiplier effect. In only one year, the 61 teacher educators taught a total of 5,000 preservice education students about gender equity in mathematics, science, and technology. If we estimate that each new teacher will teach for 25 years and have 25 students in her/his class, these 5,000 alone will encourage 1,562,500 girls to persist in mathematics, science, and technology. Participants also taught a total of 5,000 colleagues, inservice teachers, parents, and others. And this is only *one* year's impact. The percentage of participants whose syllabi mentioned gender equity doubled (from 23%to 48%) while those whose syllabi specifically targeted gender equity increased sevenfold (4% to 27%). By a measure devised to assess pre/post teaching of gender equity, 85% of the participants changed in a more equitable direction, many quite substantially. In another pre/post measure, the percentage of participants who spontaneously mentioned the impact of gender equity issues on their lives increased from zero to 21%. Interestingly, our evaluation shows that the male participants were even more likely to make large pro-equity changes in their teaching than were the female participants.

Here are comments by some of the students of participanting professors of education about what they learned from *Gender Equity Right From the Start.*

> *I am startled and upset by the figures and facts I learned this semester about equity in the classroom. It is sad that women are so underrepresented in the work force, especially in math and science fields. Hopefully my knowledge of the stereotypes that*

occur so subtly in the classroom will keep me aware of how I interact with the students. I hope to have a classroom in which I will seek equity!

I decided to watch my Field Experience teacher one morning. She had a competition of boys against girls where she asked mini-math questions. While she was fair in that each team got a turn to answer, the questions she asked were extremely biased. The boys' questions were all about sports, while the girls' were about dresses, etc. I'm sure my teacher didn't even realize it was happening. Hopefully, studying this issue and being more aware of it will prevent me from doing it to my students in the future.

The gender equity project was one of my favorite assignments in this class.

I have learned that the main way to insure gender equity in the classroom is to make a conscious effort to do so. This means making it a point to call on girls, encourage them in math and science, and challenge them to go beyond concepts to discover new questions and answers. It means preventing yourself from showing favoritism to boys.

I feel that the knowledge I have about equality in the classroom will definitely help me to be a better teacher.

Women Are Not All the Same

Source materials in this book refer to girls and women, but this is merely a convenient shorthand. There is no behavior or attitude shared by all girls or all women. Female people vary enormously in all characteristics, most definitely including learning styles.

It is also essential to remember that women come in all colors. Bulletin board displays need to show women of color. In class discussions on gender equity, consider whether there are any special implications for women of color. In field assignments concerning gender equity, pay attention not only to gender in your classroom but also to racial/ethnic groups. It does no good to advance White girls in mathematics, science, and technology while ignoring the needs of girls of color.

We need in all honesty to point out that a lot of the research on gender in education has been done without disaggregating results by racial/ethnic group in addition to sex. Another limitation of existing research is that much of it is based on middle-class girls to the exclusion of working-class groups. We have been obliged to report on what is available. Perhaps you can produce publishable-quality research that will help to right this imbalance.

Remember the Gentlemen

Many people believe that gender bias exists because men perpetuate it. Consequently, if more women were in positions of authority, influence, and power we could eradicate gender bias immediately. Right?

Wrong. There are many gender-fair men and many gender-biased women. We all began learning sexism unconsciously with the pink and blue receiving blankets we were wrapped in as newborns, and kept on learning it as children at home (think of toys for boys and toys for girls), at school, and in our communities. As we have pointed out, the men in our project did especially well. Two notable recent studies seem to support what we found by disproving the male = evil and female = virtue equations. Canes and Rosen (1995) found "no evidence that an increase in the share of women on a [science and engineering] department's faculty led to an increase in its share of female majors." [2] A project conducted at Harvard found that women in science who had female advisers during their postdocs left science at a higher rate than those with male advisers.[3] Of course, there are many other indications that women have positive role-model effects on female students as well, which seems to mean that it doesn't matter who encourages girls to persist in MST.

This is as it should be. Gender equity is not a *female* issue but a *human* issue. Most of us, whether we are men or women, have mothers, aunts, sisters, daughters, nieces, or granddaughters whose futures we care about deeply. Their lives affect ours. Considering our role as educators, this is also as it should be. We teach with our heads, hearts, and hands, not with one type or another of a reproductive system. Our shared goal, male and female educators alike, is to help the next generation achieve the most productive and satisfying lives they can — not some of them, but all of them.

Another consideration about males is that gender equity is for boys, too. It does no good to help girls build confidence and achievement in mathematics, science, and technology if boys grow up believing that these are men's fields. Boys need to become men who will be secure mentors, colleagues, supervisors, and employees of women in MST. Attention to gender equity now will help that happen.

What the Statistics Say

Several tables showing recent data on women and girls in mathematics, science, and technology education and employment are presented on the next few pages.

[2] Canes, Brandice J. and Rosen, Harvey S. (1995) "Following in her footsteps? Faculty gender composition and women's choices of college majors." *Industrial and Labor Relations Review*, vol. 48, no. 3, April issue, pages 486-504.

[3] Babco, Eleanor L., ed. (1996). "Gender disparity." *CPSTComments.* Washington, DC: Commission on Professionals in Science and Technology, January-February issue, pages 26-27.

WOMEN IN SCIENCE, MATHEMATICS, AND TECHNOLOGY EMPLOYMENT: PARTICIPATION AND EARNINGS, 1996

Occupation	*% Female*	*Annual Earnings **
Occupations Requiring a College or Advanced Degree		
Engineers	9	$49,300
Mathematical and computer scientists	31	45,900
Chemists	31	44,300
Biologists	37	36,300
Physicians (salaried)	31	58,900
Pharmacists	39	51,600
Registered nurses	91	$36,200
Dieticians	90	24,900
Pre-kindergarten and kindergarten teachers	98	18,800
Elementary school teachers	83	34,400
Secondary school teachers	55	36,200
Librarians	83	34,300
Social workers	68	27,200
Occupations Requiring an Associate Degree		
Electrical and electronic technicians	12	$31,700
Drafters	21	31,300
Surveying and mapping technicians	9	24,000
Biologic technicians	59	25,200
Chemical technicians	27	31,100
Computer programmers	30	40,100
Clinical lab technicians	71	$27,000
Licensed practical nurses	95	24,300
Dental assistants	98	18,800
Legal assistants	85	28,500
Occupations Requiring a High School Diploma or Apprenticeship		
Mechanics and repairers	4	$29,100
Electrical and electronic equipment repairers	12	33,500
Machinists	7	28,200
Sales workers, radio, TV, hi-fi, & appliances	21	22,000
Sales workers, apparel	75	$13,800
Secretaries	99	21,100
Hairdressers and cosmetologists	88	15,200
Bank tellers	91	16,400

* Calculated from weekly figures and rounded off to nearest $100

Source: Household Data, 1996 Annual Averages. Bureau of Labor Statistics, U.S. Department of Labor. To update statistics, contact BLS, (202) 606-6378 or check the Web at <http://stats.bls.gov>.

PERCENTAGE OF HIGH SCHOOL GRADUATES TAKING MATHEMATICS AND SCIENCE COURSES, BY SEX, 1994

	% Boys	*% Girls*
MATHEMATICS		
Algebra 1	64.7	68.1
Geometry	68.3	72.4
Algebra 2	55.4	61.6
Trigonometry	16.6	17.1
Analysis/pre-calculus	16.3	18.2
Calculus	9.4	9.1

(In 1982, boys took more math than girls in all courses except algebra 1 and geometry.)

SCIENCE		
Biology	92.3	94.7
Chemistry	53.2	58.7
Physics	26.9	22.0

(In 1982, boys took more science than girls in chemistry as well as physics. Then, 18.8% of boys took physics while 10.0% of girls did.)

Source: National Center for Education Statistics (1996). *The Condition of Education 1996.* Washington, D.C.: U.S. Department of Education, page 100.

----- *Compare the high school data with the postsecondary data.* -----

PERCENTAGE OF POSTSECONDARY DEGREES AWARDED TO WOMEN, 1992-1994

Field of Study	*Associate*	*Bachelors*	*Masters*	*Doctorate*
Mathematics	38	46	38	22
Biological/life sciences	59	51	55	41
Physical sciences, total	42	34	29	22
Chemistry	na	41	41	28
Physics	na	18	15	12
Computer & information scis.	51	28	26	15
Engineering	13	16	15	11

Source: National Center for Education Statistics (1996). *Digest of Education Statistics1996.* Washington, DC: U.S. Department of Education, Tables 241 and 244. Associate degree data for 1992; data for other degrees is 1994.

EDUCATION EMPLOYMENT

	Percent Women
Pre-K & kindergarten teachers (1)	98
Elementary teachers (1)	83
Secondary teachers (1)	55
Public school principals (2)	35
Superintendents (3)	7
College and university professors, full time (4)	33
College and university professors, part time (4)	45

Data Sources:

1 Bureau of Labor Statistics, U.S. Department of Labor. 1996 data.
2 National Center for Education Statistics (1996). *1996 Digest of Education Statistics.* Washington, DC: U.S. Department of Education, Table 86, 1994 data.
3 American Association of School Administrators, 1993 data.
4 National Center for Education Statistics (1996). *1996 Digest of Education Statistics.* Washington, DC: U.S. Department of Education, Table 226. 1992 data.

WOMEN EMPLOYED FULL-TIME IN COLLEGES AND UNIVERSITIES BY RACE/ETHNICITY, 1992

Field	*Total Employed Men and Women*	*% Women in Field*	*% White Women*	*% Black Women*	*% Hisp. Women*	*% Asian-Amer. Women*	*% Amer. Indian Women*
Engineering	24,431	6	3.9	0.7	0.2	1.3	—
Biological sciences	34,289	23	20.8	1.5	0.5	0.9	0.1
Physical sciences	28,313	12	10.2	0.3	0.1	0.9	—
Mathematics	25,325	25	21.6	1.0	0.6	1.2	0.4
Computer sciences	13,578	20	17.8	1.3	0.1	0.8	—
Teacher education	12,490	57	50.3	4.9	0.7	1.0	0.2

Source: National Center for Education Statistics (1996). *1996 Digest of Education Statistics.* Washington, DC: U.S. Department of Education, Table 227, 1992 data.

SOURCE MATERIALS

EQUITY ISSUES

OVERVIEW

GENDER EQUITY IN MATHEMATICS, SCIENCE, AND TECHNOLOGY EDUCATION

Gender differences in MST performance and participation are a function of contexts or situations in which these subjects are taught and learned. In order to become familiar with the dynamics of teaching them effectively and equitably to females, preservice teachers should be aware of the research defining the problem and giving recommendations on dealing with gender bias in the classroom.

EQUITY ISSUE 1

MATHEMATICS, SCIENCE, AND TECHNOLOGY AS MALE DOMAINS

Mathematics, science and technology are perceived as male domains, from which women are thought to be absent. This erroneous belief nevertheless creates its own reality when girls and women come to feel the fields are inappropriate for them and act on it by failing to pursue MST courses beyond minimum requirements.

1A. History of Women in MST

The apparent absence of women from many histories of achievement in MST perpetuates the myth that these fields have always been and thus are "naturally" male domains. As Sara Evans of the University of Minnesota states, "Having a history is a prerequisite to claiming a right to shape the future." It is important to demonstrate to students that women have indeed made essential contributions.

1B. Lack of Women Inventors In Curriculum

The lack of women inventors in texts and other standard curriculum materials gives the erroneous impression that women did not invent anything. Learning about women inventors dispels this myth.

1C. Invisibility of Contemporary Role Models

The invisibility of women as contemporary scientists, mathematicians, and technologists generates and reinforces girls' belief that these fields are male activities that do not welcome women's participation. Without role models, it is hard for girls to envision themselves as potential specialists in these fields, and many opt out of continued study.

1D. Reinforcement of Male Stereotypes in Media

When media — movies, television, radio, newspapers, magazines, books, computer software, and others — show the vast majority of mathematicians, scientists, and technologists as male, girls (and boys) understand this as descriptive of reality. It is a small step to prescriptive: It is "normal" and "natural" for these people to be male. This becomes a self-fulfilling prophecy.

1E. THE CULTURE OF MST FAVORS MEN

The culture — the mores, environments, ways of doing things, systems of reward and punishment — of mathematics, science and technology is one that reflects its male history and therefore has many traditionally male features: competition over cooperation, hierarchy, aggressiveness, long hours that exclude family time, and others. Women and girls may feel uncomfortable and unwelcome in this culture, which can however be altered with the presence of a sufficient number of women and men willing to consider new ways.

1F. ISOLATION OF GIRLS AND WOMEN IN MST

While sociologists of science and related fields assert that the meritocracy works, stories from the field by women indicate that isolation within the community of male scientists is prevalent. Females entering MST fields need to develop strategies for dealing with this marginalization.

EQUITY ISSUE 2

PEERS', TEACHERS', PARENTS', AND SOCIETY'S CULTURAL EXPECTATIONS

The notion that males excel in mathematics, science, and technology is one of many beliefs and cultural influences that are passed down from generation to generation. Teachers, parents and others transmit signals to children about who can succeed in MST. The dynamic is all the more powerful in that they themselves may not realize they hold these beliefs and act on them. The subtle and unintended messages can create the belief among girls that they cannot be successful in these fields. When children perceive this attitude in adults, children reflect and reinforce it through their interactions with their peers. This cultural expectation becomes a self-fulfilling prophecy — a cycle that continues to discourage women's participation in MST.

2A. BIASED BELIEFS ABOUT WHO SUCCEEDS IN MST

When teachers, parents, and peers believe that males will perform better in mathematics, science, and technology than females, many girls internalize this belief and behave accordingly. When parents' biased beliefs lead them to feel that these fields are inappropriate choices for a girl, they can discourage their daughters from persisting in them, either actively or by failing to encourage them. It is important to identify gender-based beliefs that have no basis in reality in order to dispel them.

2B. CHOICE OF MATERIALS BY TEACHERS AND PARENTS REINFORCE SEX STEREOTYPES AND PRODUCE NEGATIVE ASSOCIATIONS

The choice of materials such as toys, books, and clothing reinforces cultural expectations of female behavior. Girls who are not exposed to toys and materials that encourage scientific, mathematical or technological thinking and activities are less likely to develop an interest in these subjects, thus reinforcing the belief that they are not for girls. Similarly, when girls are exposed to materials negatively portraying women in these areas, they are discouraged from maintaining their interest in them.

2C. Cultural and Parental Influences Can Work Against MST for Girls

In some cultures, parents discourage females from pursuing fields in science, mathematics, or technology. This can create problems for teachers who encourage female students, and conflicts for the girls who find these subjects interesting and exciting. Preservice teachers need to be sensitive to parents' attitudes, beliefs, and cultural issues about who succeeds in MST.

2D. Lack of Awareness of Gender Bias

Gender biased beliefs are both taught and expressed by the subtle and unintentional behaviors of parents, teachers, and peers. Preservice teachers who are gender biased, most definitely including young women, will pass this bias on to their students the same way it was passed to them. Preservice teachers need to identify their own bias and investigate ways of counteracting biased behavior.

2E. Lack of Techniques to Counteract Bias

Once teachers become aware of the problem of gender bias, what are they supposed to do about it? When when they don't know, gender imbalances continue unchecked. Becoming aware of the issues is an important first step, but without knowledge of what to do awareness is futile. The next step in breaking the cycle of bias is to learn about proven techniques used in early intervention or outreach programs that counteract bias and inequity, and strategies that work to address biased cultural expectations for females in mathematics.

EQUITY ISSUE 3

BIASED AND INAPPROPRIATE CURRICULUM MATERIALS

Curriculum materials that are male-biased in language, content, and/or illustrations reinforce for females the idea that mathematics, science, and technology are male domains. The situation is further complicated by materials that focus on abstractions and procedure-based learning, which only reinforces beliefs that these subjects are impersonal and irrelevant to real life. This is especially problematic for females who, by using biased curriculum materials, will see MST as unrelated to their daily life experiences in and out of school and will not be able to envision themselves using these skills in their adult lives as citizens or workers.

3A. Curriculum Materials Show A Predominance Of Males

When curriculum and classroom materials present notions of male presence and female absence or male activity and female passivity, girls are reinforced in a belief that MST is a male activity. It is important to identify any male bias in curriculum materials and compensate for it.

3B. WOMEN ARE SHOWN IN TRADITIONAL ROLES

Males and females are often assigned traditional roles in the materials that students are exposed to, reinforcing the idea that MST are male fields and that women are expected to be helpers, at best. Similarly biased representations in print and software materials often go unnoticed, especially because they are so familiar, and the biases are communicated to children. Reviewing materials with an eye to such bias helps preservice teachers develop awareness of the issues and be more discriminating in their selection of materials.

3C. MATERIALS DO NOT REFLECT GIRLS' INTERESTS

Many times girls lose interest in mathematics, science, and technology because they find the content based on interests frequently held by males and not females. While it is important to avoid reaching out to girls exclusively through the traditionally limited areas that they have been encouraged to engage in such as cooking and sewing, it is important that preservice teachers find ways to involve girls in MST through content that truly interests them.

3D. MATERIALS DO NOT REFLECT THE REAL-LIFE RELEVANCE OF MST

Making the connections of MST concepts to everyday life, such as a practical use for a scientific discovery or a mathematical algorithm, or understanding everyday phenomena, fosters the interest and participation of girls much more than materials that do not make connections and deal with depersonalized, abstract concepts.

3E. BIASED LANGUAGE IN CURRICULAR MATERIALS

Our choice of words conveys our biases and assumptions about the world around us, and the idea that mathematics, science, and technology are for males and not for females is embedded in many curriculum materials, especially older ones. When we accept the language, we accept the idea as well. Preservice teachers need to become aware of language biases as they review and select materials for their classes.

3F. MATERIALS FOSTER ISOLATED, COMPETITIVE ACTIVITIES

Research shows that girls often prefer working in collaboration with others as opposed to working individually or in competition with others. When curriculum materials foster isolated, competitive work, girls are more likely to lose interest. This can influence their decisions to continue working with and learning about mathematics, science and technology. Preservice teachers need to identify these aspects of learning materials and find alternatives.

EQUITY ISSUE 4

CLASSROOM INTERACTION AND ATMOSPHERE

When teachers have biased beliefs, research has repeatedly and conclusively shown that they inadvertently express them to their students in the form of biased behaviors. This is true of both women and men. Small and often subtle behaviors favoring males by teachers and — permitted by teachers — classmates, which can include a competitive or aggressive classroom atmosphere, serve to discourage girls and young women from continuing in mathematics, science

and technology. While each individual incident may be trivial, the accumulation of biased incidents is powerful. Often girls themselves do not realize the source of their discouragement. Only by identifying problem behaviors can they be eliminated.

4A. BIASED TEACHER/STUDENT INTERACTIONS

Many teachers inadvertently favor boys in their classroom behavior. Research shows that especially in male-dominated subjects such as mathematics, science, and technology, teachers call on boys more often than girls, wait longer for boys' answers, and engage boys when they call out, while reprimanding girls and reminding them to raise their hands. Research also shows that when teachers receive training in unbiased classroom behaviors and feedback on their own behavior, these imbalances can be corrected. Preservice teachers need to examine their own behavior before they begin their teaching careers.

4B. PHYSICAL ENVIRONMENT

The physical design and affective climate of the classroom can influence collaborative or isolated student activities. How a classroom is set up in terms of student groupings, accessibility of equipment and materials, bulletin boards, and seating arrangements can subtly encourage or discourage participation in classroom activities. Because group work and peer teaching are conducive to greater female participation, arranging a classroom to favor these activities encourages the creation of more "female-friendly" interactive environments.

4C. PEER AGGRESSIVENESS

Research has shown that males receive more attention and therefore more help during MST lessons in part because they show more aggressive behaviors than females. Aggressiveness is culturally sanctioned behavior for boys —"Boys will be boys!" — and can often be physical in elementary grades and verbal in secondary grades. Commonly it shows up when teachers fail to stop boys (or occasionally girls) who make aggressive or hostile comments or exhibit such behaviors to female classmates or quieter male classmates. When girls don't feel safe from ridicule in expressing their opinions or answering questions, they learn not to venture opinions or answers in class. A silent, withdrawn student is likely to believe that she is not very good at mathematics, science, or technology, which can become a self-fulfilling prophecy. Preservice teachers need to be able to identify and deal constructively with aggressive male behavior and be aware of behaviors that will encourage female success.

4D. INSTRUCTIONAL STRATEGIES DO NOT FOSTER COOPERATIVE LEARNING AND COLLABORATION

Research shows that many, if not most, females, as well as a sizeable proportion of males, learn best in a cooperative, social, and collaborative environment. This is of course the environment in which most scientists, mathematicians, and technologists actually work. To the extent that the educational process or environment reflects isolated or competitive modes of learning and relating, we can expect that a number of girls will be unnecessarily disadvantaged. There are many ways that cooperative learning and collaborations can be used in the teaching and learning of mathematics, science, and technology.

EQUITY ISSUE 5

ANTI-INTELLECTUALISM AND ATTRIBUTIONAL STYLE

One reason gender differences exist in performance and participation in mathematics, science, and technology is because of the lack of females' confidence when doing them. Achievement motivation, attributional style, and pressures from society can contribute to a lack of confidence and a fear of success. Some teaching behaviors toward girls and women can encourage a learned helplessness that further reduces confidence and eventually results in the loss of self-esteem. Traditional female characteristics of helplessness and risk avoidance are incompatible with forthright intellectual achievement in MST, and achievement is considered a fluke rather than earned. This belief is unfortunately strengthened by anti-intellectual currents in mainstream American society. Preservice teachers need to recognize these influences and learn how to counteract them.

5A. ATTRIBUTIONAL STYLE: GIRLS ARE LUCKY AND BOYS ARE SMART

Research shows that there tends to be a difference in how boys and girls interpret their successes and failures. Boys typically attribute their successes to their own intelligence and their failures to insufficient effort. Many girls believe their successes are due to luck and their failures to inability. So when girls fail, there is no motivation to try again: Trying harder or risking a new approach won't make much difference if you're simply not very good at the subject at hand. Girls need to know that they can apply their intelligence effectively in the area of math, science, and technology, or they will be discouraged from continuing.

5B. LEARNED HELPLESSNESS

While boys are often encouraged to investigate possible approaches to problems and obstacles, teachers, who do not realize they hold different expectations and treat children differently by gender, often finish tasks for girls who hit a road block. This is one of the ways in which girls learn to believe they are not capable of finding solutions or solving problems on their own. Girls need to have experiences with developing their own autonomous problem-solving abilities or they will have difficulty becoming independent learners and capable adult workers. Preservice teachers need to realize that "doing it for the girls" instead of coaching them to do it for themselves does them no favors.

5C. RISK-TAKING BEHAVIOR

Mathematics, science, and technology are learned best when students are encouraged to construct their own knowledge, which requires taking risks, trial and error, and learning from mistakes. This type of behavior is often not fostered in females, who can feel incompetent when they do not get the correct answer immediately. Preservice teachers need to develop a repertoire of instructional techniques that create situations where females can construct knowledge, take risks, tolerate ambiguity, and investigate solutions in order to develop an understanding and appreciation for MST.

5D. SOCIAL PRESSURES

Social pressures on children to behave in stereotyped sex-role ways increase drastically as they enter adolescence. Research shows that girls' self-esteem drops dramatically as they enter the upper grades and social issues take priority over academic ones. Because girls think they will be unpopular if they are perceived as brainy or if they perform better than boys in MST, many succumb to the pressure to be "feminine," which means not being too smart and not being too good at "boy stuff" — science, math, and technology. The conflicts are further complicated for women from cultures where MST are considered unnecessary or even unfeminine for girls. Preservice teachers need to learn techniques that support a learning environment where learning MST is a positive experience for girls and where their success is respected.

EQUITY ISSUE 6

TESTING AND ASSESSMENT

Traditional forms of testing can place girls at a disadvantage when methods and content capitalize on skills boys are more likely to have than girls. We know that females are better course takers than they are test takers. Gender differences in achievement may be the result of the testing process and not an indication of girls' lesser abilities. They moreover perpetuate stereotypes of girls' "deficient" achievement in mathematics, science, and technology. The recent advent of performance assessment and more authentic measurements for assessing what students know in real-life contexts is consonant with what works best for girls and young women.

6A. TEST DESIGN FAVORS MALES

Girls have historically scored lower on MST achievement tests than do boys. This may be due to bias in testing methods, not to ability. Girls are less likely to take a risk and guess at multiple-choice questions than are boys. They do less well in competitive timed test situations and when tests require decontextualized bits of information that have been memorized. Another issue is the way test results are reported, including the bell curve. Exposing preservice teachers to what we know about girls and testing and alternative modes of assessment will help encourage girls' achievement in science.

6B. TEST ITEMS FAVOR MALE INTERESTS

Research has shown that boys tend to do better than girls with test items whose content and context favor male interests, refer less to women, or present women in stereotypical roles. Girls tend to have more success with real-life application problems in mathematics, science and technology, and with process skill questions than straight content questions. When test content does not reflect girls' knowledge and strengths, their scores are artificially low.

STARTLING STATEMENTS — 1994-1995

EQUALS
Lawrence Hall of Science, University of California, Berkeley, CA 94720
Fax (510) 643-5757, Phone (510) 642-1823

1. Forty-five percent of the high school juniors who take the PSAT are boys. What percent of the ultimate scholarship winners are boys?

2. In 1991, 15% of families with children were denied shelter. What percent were denied shelter in 1993?

3. What percent of 700 school districts surveyed nationwide reported at least one shooting or knifing in school?

4. In 1983, the proportion of the U.S. Latina/o population 25 years or older with at least a high school diploma was 46%. What was it in 1993?

5. What percent of U.S. residents over 5 years of age speak a language other than English at home?

6. What percent of states require coursework for elementary teachers in parental-involvement techniques?

7. The U.S. General Accounting Office surveyed nearly 500 school- or community-based dropout programs. What percent served Latina/o youths?

8. What is the median annual earnings figure for African-American females in the U.S.?

9. What percent of school superintendents are women?

10. About 24% of U.S. students complete four or more years of college, whether they are male or female. In Japan about 34% of males do so. What percent of Japanese females complete higher education?

11. About what percent of the completed suicides are committed by lesbian and gay youth annually?

12. What percent of United States women work full time?

13. The National Council of Teachers of Mathematics (1989) *Curriculum and Evaluation Standards for School Mathematics* states, "all students will have a calculator." What percent of 8th grade students reported having unrestricted use of calculators in their math classes?

14. What percent of African Americans who graduated from high school, enrolled in college the following fall?

15. The U.S. has one Asian-American senator. How many Asian Americans are members of the House of Representatives?

16. The National Council of Teachers of Mathematics (1989) *Curriculum and Evaluation Standards for School Mathematics* states, "A computer will be available at all times in every classroom." What percent of 12th grade students had at least one computer in their mathematics classroom?

STARTLING STATEMENTS (1994-1995) — ANSWERS

1. 60% American Civil Liberties Union Women's Rights Project on Behalf of the National Center for Fair and Open Testing, Cambridge MA, 1994.

2. 29% Survey of Homelessness, Conference of Mayors, Washington, DC, December 1993.

3. 39% Survey by National School Boards Association, Washington, DC, January 1994.

4. 53% U.S. Census Bureau Report, 1993, on the Hispanic Population as reported in *Hispanic*, August 1994.

5. 14% U.S. Census Bureau as quoted in *Education Week*, vol. XII, no. 33, May 12, 1993.

6. 6% *Report on Parental Involvement Training*, Center for School Change, Hubert H. Humphrey Institute, University of Minnesota, Minneapolis, MN, 1994.

7. 5% U.S. General Accounting Office as quoted in *Education Week*, vol. XIII, no. 37, June 8, 1994.

8. $20,299 U.S. Department of Commerce, Bureau of the Census, *Money Income of Households, Families and Persons in the United States: 1992*, Washington, DC, September 1993.

9. 3% *Course Trends*, July-August 1994.

10. 12% *Indicator of the Month*, National Center for Education Statistics, Office of Educational Research and Improvement, U.S. Department of Education, December 1993.

11. 30% FactFile, Hetrick-Martin Institute, New York, NY, 1992. Anderson, J. "School Climate for Gay and Lesbian Students and Staff Members," *Phi Delta Kappan*, vol. 76, no. 2, October 1994.

12. 55% Vetter, B. *Scientific, Engineering, Technical Manpower Comments*, vol. 31, no. 4, June 1994, p. 15.

13. 19% Council of Chief State School Officers, *State Indicators of Science and Mathematics Education*, 1993, Washington, DC, p. 4.

14. 49% Vetter, B. *Scientific, Engineering, Technical Manpower Comments*, vol. 30, no. 4, June 1993.

15. 6 Congressional Quarterly, *Weekly Report*, January 9 1993.

16. 20% National Center for Education Statistics, U.S. Department of Education, 1992. National Council of Teachers of Mathematics (1989), *Curriculum and Evaluation Standards for School Mathematics*, Reston VA, p. 124.

COMMENTS ON THE 1994-1995 STARTLING STATEMENTS

1. Each year, more than a million high school juniors take the PSAT. The top 15,000 scorers, 60% of whom are boys, automatically become National Merit semifinalists. These students then submit records of their grades, extracurricular activities, recommendations, and essays; about 14,000 are chosen as finalists. National Merit Scholarship officials say 60% of the ultimate winners are boys. A *Harvard Educational Review* study of 47,000 students found that women who scored 33 points lower than men on the SAT subsequently earned the same grades in the same college math courses as their male peers.

2. According to the survey of 26 cities, 43% of the homeless population are families with children. Unaccompanied children are 3% of the homeless. Shelters in 85% of the surveyed cities had to turn away families for lack of resources. Overall, about 25% of requests for shelter were unmet.

3. In a survey of 700 school districts, the National School Boards Association found that reports of violence were highest in urban districts; however, both suburban and rural districts reported increases in the number and seriousness of violent incidents, including rapes and shootings. Seventy-seven percent of the responding school districts reported the causes as changing family situations. Sixty percent blamed portrayal of violence on television. The methods for combating the violence ranged from police presence on campus to the use of metal detectors and closed-circuit television surveillance.

4. According to a U.S. Department of Education National Center for Education Statistics survey, all of the Latina/o tenth- through twelfth-graders who stated school reasons for dropping out of high school in 1992, 64% claimed they did not like school or did not feel like they belonged. Latinas/os constitute approximately 36% of the U.S. minority population. Latinas/os are the second largest minority group in the United States, with a mainland population of 22.35 million in the 1990 Census. (Where we have used Latina/o, the term "Hispanic" is used by the U.S. Bureau of the Census to refer to Americans of Mexican, Puerto Rican, Cuban, Central and South American, and Spanish descent.)

5. A language other than English is spoken at home by 31.8 million Americans age 5 and older: That is one out of every seven. The ten most common home languages in the U.S. in order of population size are: Spanish, French, German, Italian, Chinese, Tagalog (spoken in the Philippines), Polish, Korean, Vietnamese, and Portuguese. The fastest growing language was Mon-Khmer, spoken by Cambodians. The number of Mon-Khmer speakers increased by 676% over a 10-year period.

6. Only three states — Iowa, Minnesota, and Virginia — have specific coursework requirements in parent-involvement techniques for prospective elementary teachers, and none requires secondary teachers to complete such courses. The national Council for the Accreditation of Teacher Education revised its standards to require that prospective teachers complete a sequence of courses or fieldwork preparing them to work with parents.

7. Latina/o students drop out of school at a rate nearly four times that of European Americans and a rate more than twice that of African-American students. Latina/o students drop out of school earlier than other groups. Forty percent of Latina/o dropouts ages 16 to 24 left school with less than a 9th-grade education, compared with 15% of European Americans and 10% of African-American students. Among students with low family incomes, 45% of Latinas/os were dropouts compared with 24% of African Americans and 19% of non-Latina/o Whites.

8. The median annual earnings figure for full-time working U.S. males, regardless of position or education, is $31,012. The median annual earnings figure for full-time working females is $22,167. The median annual earnings of full-time workers include:

European-American males	$31,737
African-American males	$22,942
European-American females	$22,423
Latino males	$20,312
African-American females	$20,299
Latina females	$17,743

9. Eighteen states elect their state school superintendents by popular vote; in 32 states they are appointed. Of the nine women, eight were elected and one was appointed.

10. Compared to other large industrialized countries, the United States has the most educated population overall. Young men in Japan were much more likely to have completed higher education than men in other industrialized countries. Young men in the United States ranked second. Young women in the United States were more likely to have completed higher education than women in Japan, and more likely to have completed higher education than either men or women in the other countries.

11. At least two states, Connecticut and Massachusetts, have suggested specific ways in which homophobia can be addressed in the schools.

12. Women accounted for 69% of total labor force growth with the total number of women working increasing by about one million. Median earnings for women working full time are currently 77% of the earnings of their male counterparts.

13. The percentage of students with unrestricted calculator use was positively associated with each state's average in mathematics proficiency. Only 3% of fourth graders were permitted free and open use of calculators in class. When taking tests, 2% of fourth graders and 34% of eighth graders were permitted to use calculators. Although there are some exceptions, more proficient students appear to have more opportunities to use calculators than their less proficient peers.

14. Women (65%) were more likely to enroll than men (60%). Fifty-seven percent of Latina/o Americans enrolled in the Fall, while 64% of European Americans did so.

15. There is one African-American senator and 38 African Americans in the House of Representatives. There are no Latina/o senators. There are 20 Latina/o representatives in the House. The Senate has one Native American and there are no Native Americans in the House. There are six women in the Senate and 46 in the House of Representatives.

16. About one-third of fourth-grade students had at least one computer in their mathematics classrooms, while one-fifth of eighth- and twelfth-grade students had at least one computer in their mathematics classrooms. Computer labs existed in the schools of 56% of students in fourth grade, 61% of students in eighth grade, and 79% of students in twelfth grade. These labs were not necessarily available for mathematics instruction.

DUMBER BY DEGREES

By Jo Sanders

Sexism doesn't stop at the grade school door, according to recent studies. Here's what happens as girls move through school.

1. Girls start school with higher test scores than boys. But by the time they take the SAT, boys outscore girls on both the verbal and the math exams: verbal, by 4 points; math, by 41 points.

2. In college classes, women are twice as likely as men to say nothing during class discussions.

3. After the first year of college, women show much sharper drops in self-confidence than men do. The longer women stay in school, the lower their self-confidence.

4. Women earn only 24% of doctorates in mathematics, 22% of doctorates in the physical sciences, and 9.6% of doctorates in engineering.

5. Three percent of school district superintendents are women.

6. Less than 15% of full professors are women.

1. National Report on College-Bound Seniors , College Entrance Examination Board, 1994.

2. Sadker, Myra and Sadker, David (1994). *Failing at Fairness: How America's Schools Cheat Girls*, page 170.

3. Arnold, Karen (1995). *Lives of Promise: What Becomes of High School Valedictorians.* San Francisco: Jossey-Bass Publishers.

4. Chronicle of Higher Education Almanac, September 1, 1995.

5. Course Trends, July-August 1994.

6. *Digest of Education Statistics 1995*, National Center for Education Statistics, Table 218.

FAMOUS WOMEN MATHEMATICIANS

Hypatia	370-415 AD. Greek mathematician, inventor, philospher, teacher, and textbook author.
Maria Agnesi	1718-1799. Professor at the University of Bologna in Italy, she published *Analytical Institutions* ,which was widely translated and used as a textbook.
Sophie Germaine	1776-1831. Researcher in number theorie and mathematical physics, winner of grand prize in French Academy of Science contest.
Mary Somerville	1780-1872. Popular science writer.
Ada Byron Lovelace	1815-1852. "Mother" of computer programming.
Sonya Kovaleskaya	1850-1891. Researcher in applied mathematics, recipient of the Prix Bordin from the French Academy of Science.
Grace Chisolm Young	1868-1944. First woman to receive a formal doctorate in any subject in Europe; geometer, textbook author.
Emmy Noether	1882-1935. "Mother" of modern algebra. Professor of mathematics at Bryn Mawr College and a member of the Institute for Advanced Study at Princeton, she discovered many methods of great importance to abstract algebra.
Marjorie Lee Browne	1914-1979. One of the first two Black women to receive a doctorate in mathematics in the U.S.
Cathleen Morawetz	1923- . First American women to head a mathematics institute, in her case the Courant Institute of Mathematics.

Jo Sanders, 1994

FAMOUS WOMEN MATHEMATICIANS, SCIENTISTS AND TECHNOLOGISTS

Elizabeth Blackwell, 1821-1910	**Physician.** After she became the first woman doctor in the United States, no hospital would admit her. She then bought her own house and established a small dispensary which expanded to become the New York Informary for Women and Children.
Mary Bunting, 1910-	**Microbiologist.** She is known for discoveries on the effect of radiation on bacteria and became the first woman member of the U.S. Atomic Energy Commission.
Eleanor Margaret Burbidge, 1919-	**Astronomer.** As the first woman Royal Astronomer, she directed the Royal Greenwich Observatory in England.
Annie Jump Cannon, 1863-1941	**Astronomer.** Called the "census taker of the sky," she classified about 400,00 stellar bodies according to their temperature.
Rachel Carson, 1907-1964	**Marine Biologist.** Her writings brought to public attention the destructive effects of pesticides, resulting in curtailment of their use, and in stronger efforts to develop natural biological controls of harmful insects.
Eugenie Clark, 1992-	**Marine Biologist.** Through studying the nature and behavior of sharks, she had been able to teach them to choose between targets of different designs and colors.
Jewel Plummer Cobb, 1924-	**Cell Physiologist.** Her early research led to new discoveries concerning normal and malignant pigment cells. Interested in helping to solve the puzzle of cancer, she undertook many research projects for the National Cancer Institute. In 1981, she became the first Black women college president in the 19-college California State University system.
Anna Comena, 1083-1148	**Physician, Mathematician.** Author of a book on gout, she also ran a hospital of 10,000 beds and provided medical services to pilgrims, and Crusaders and their camp-followers.
Gerty Cori, 1896-1957	**Biochemist, Physician.** The first women to receive the Nobel Prize for Medicine/Physiology, she carried out research into carbohydrate metabolism, the method by which the body uses its fuel supply of starches and sugars, and the relation of this mechanism to certain hormone secretions.
Marie Curie, 1867-1934	**Chemist.** The first person ever to receive two Nobel awards, Physics in 1903, and Chemistry in 1911, she won the prizes for her discoveries of radium and her researches into radioactivity.
Amelia Earhart, 1898-1937	**Aviator.** The first women to fly solo across the Atlantic, she set many aviation records, including long distance ones, such as that of flying from Hawaii to Calfornia.

Lillian Gilbreth, 1878-1972	**Industrial Engineer.** She charted ways to save energy and human motion in the office, factory, hospital, and home. She also had 12 children!
Maria Goeppert-Mayer, 1906-1972	**Physicist.** She received the Nobel Prize for Physics in 1963; during World War II, she worked on isotope separation for the atomic bomb.
Jane Goodall, 1934-	**Animal Behaviorist.** Living for long periods of time in the field, her studies were the first to show that chimpanzees are intelligent, social animals.
Beatrice Hicks, 1919-	**Electrical Engineer.** In 1960s she was the only Black woman engineer at Western Electric. She received the Society of Women Engineers Achievement Award for her theoretical study and analysis of sensing devices under extreme environmental conditions. She is a graduate of Newark College of Engineering (now N.J.I.T.).
Lucy Hobbs, 1833-1910	**Dentist.** The first women to earn a dental degree in the United States, she developed one of the most extensive practices in Kansas.
Dorothy Crowfoot Hodgkin, 1910-	**Crystallographer.** She was awarded the Nobel Chemistry Prize in 1964 for determining the crystal structure of biomedical compounds, particularly penicillin.
Grace Hurray Hopper, 1906-	**U.S. Naval Officer, Inventor.** She is the inventor of the computer language COBOL (Common Business-Oriented Language), for which she received the 1983 American Association of University Women (AAUW) Educational Foundation Achievement Award.
Shirley Jackson, 1946-	**Physicist.** The first Black women to graduate in theoretical physics from Massachusetts Institute of Technology, she works at Bell Laboratories using computers and mathematical formulas to explore physics. She was a visiting scientist at the European Organization for Nuclear Research in Switzerland.
Irene Joliot-Curie, 1897-1956	**Physicist.** She was awarded the Nobel Chemistry Prize for discovering a technique for making artificial radioactive elements.
Reathea Clark-King, 1938-	**Research Chemist.** One of only two Black graduate students at the time, she earned her master and doctorate degrees in chemistry at the University of Chicago, specializing in high temperature chemistry. She later became a college administrator, supporting programs for women and minorities.
Barbara McClintock, 1902-	**Biochemist.** She received the 1983 Nobel Prize for Medicine/ Physiology for her discovery that genes can move from one spot to another on the chromosomes of a plant, thus changing future generations of plants.

Elsie Gregory MacGill, 1905-	**Aeronautical Engineer.** She was the first woman to become chief aeronautical engineer of any company and is known internationally for her work on engineering designs of fighter and transport aircraft.
Margaret Mead, 1901-1978	**Anthropologist.** She was one of the first to do field work in the island of the southwest Pacific and to bring back eyewitness accounts of the native cultures and peoples of New Guinea, Samoa, and Fiji.
Lise Meitner, 1878-1968	**Physicist.** The first woman to be awarded the Enrico Fermi Award, she worked on splitting the atom, and explained mathematically the fission on the uranium atom.
Maria Mitchell, 1818-1899	**Astronomer.** Self-taught, she discovered a comet that was named for her, and became professor of astronomy at Vassar College.
Jennie R. Patrick, 1949-	**Chemical Engineer.** The first Black woman in the United Stated to earn a doctoral degree in chemical engineering, she does research on energy conservation and pollution control. As a role model, she encourages young people to develop their own potential and establish their own goals.
Susan Laflesche Picotte, 1865-1915	**Physician.** Born to the Omaha Reservation, this Native American woman attended the Woman's Medical College of Pennsylvania, and returned to serve her people's medical needs. She later became the leader of the Omahas and represented their interests in Washington.
Sally Kristen Ride, 1951-	**Astronaut.** Graduate of Stanford Unviersity with a PhD in physics, she was the first American woman to go into space aboard the space shuttle Challenger as flight engeineer for nearly six days.
Florence Sabin, 1871-1953	**Physician.** She was one of the first women to enter medical research. She determined the origin of red corpuscles and did important research on tuberculosis.
Margaret Sanger,1883-1966	**Public Health Nurse.** A pioneer in birth control, she fought for the right of women to have access to birth control and was one of the founders of Planned Parenthood.
Susan Smith McKinney Steward, 1847-1918	**Physician.** She was the first Black woman to receive her MD from the New York Medical College for Women. She helped to found a hospital for treatment of "indisposed" shop girls.
Ellen Swallow, 1842-1911	**Chemist.** The first woman to graduate from the Massachusetts Institute of Technology, she was the founder of home economics as a science and a profession.
Valentina Tereshkova, 1937-	**Cosmonaut.** A Russian textile worker, she learned to parachute and eventually entered the space program. She is the first women to orbit the eath in space.

Adah Belle Thoms, 1863-1943	**Nurse.** She was the first Black person to hold an administrative position at New York City's Lincoln School for Nurses. She played an important role in guiding the progress of Black women in the field of nursing.
Jane C. Wright, 1919-	**Cancer Researcher.** Motivated by the challenge to find a cure for cancer and to help cancer patients live more comfortable lives, she has worked in chemotherapy research since 1949.
Chien-Shiung Wu, 1912-	**Physicist.** She made important contributions to the research of nuclear forces and structures. In particular she helped to prove the principle of parity unacceptable.
Rosalyn S. Yalow, 1921-	**Nuclear Physicist.** She was awarded the Nobel Prize for Medicine in 1977 for the discovery of radioimmunoassay, a method of measuring minute concentrations of hundreds of substances in body tissues. It is invaluable in determining the differences between diseased and normal states of body tissues.

Marylin A. Hulme, Consortium for Educational Equity,
Rutgers University, 1987

WOMEN INVENTORS

Invention	Inventor	Year
Frozen food packaging that can be heated in a microwave	Lynn Deffenbaugh	1987
A process of growing human skin in the lab for burn patients	Gail Naughton	1990
The illusion transmitter, which sends 3-D satellite images	Valerie Thomas	1980
Scotchgard, which enables fibers to resist dirt and stains	Patsy Sherman	1974, 1977
Double-pointed iron with detachable insulated handle	Mary Florence Potts	1876
The Snugli baby carrier	Ann Moore	1969
A combined plow and harrow	Anna Trexler	1888
An electrical-powered permanent waving machine for hair	Marjorie Joyner	1928
An improved fountain pen	Susan E. Taylor	1858
A machine for knitting Argyle socks	Anne McDonald	1985
A fruit press: A mechanical juicer for California oranges	Madeline Turner	1916
A machine that folds square bottoms to enable paper bags to stand by themselves	Margaret Knight	1879
The windshield wiper	Mary Anderson	1903
A home security system that utilizes television surveillance and audio intercom	Marie Van Brittan Brown	1969
Stove-Top Stuffing, an instant stuffing mix	Ruth Siems	1975

Source: National Women's History Project, "Inventive Women" poster series, 1993.

WOMEN NOBEL PRIZE WINNERS

1903	Physics	Marie Sklodowska Curie
1911	Chemistry	Marie Sklodowska Curie
1935	Chemistry	Irene Joliot-Curie
1947	Biochemistry	Gerty Radnitz Cori
1963	Physics	Maria Goeppert-Mayer
1964	Chemistry	Dorothy Crowfoot Hodgkin
1977	Medical Physics	Rosalyn Sussman Yalow
1983	Medicine / Physiology	Barbara McClintock
1986	Medicine / Physiology	Rita Levi-Montalcini
1988	Biochemistry	Gertrude B. Elion
1995	Medicine / Physiology	Christiane Nusslein-Volhard

LIFE AS A MEMBER OF THE OPPOSITE SEX

Imagine waking up tomorrow morning as a person of the opposite sex. Consider the following questions:

1. How would your behavior change? Would you be more active or less? More outspoken or less?

2. How would your clothing change? Would you be more comfortable or less?

3. Would other people treat you differently? If so, in what ways?

4. Would your students treat you differently? How?

5. Would you be more or less involved with mathematics, science or technology? In what ways?

"My Daddy Might Have Loved Me": Student Perceptions of Differences Between Being Male and Being Female

by Alice Baumgartner

Below are excerpts from a 1982 report on a survey designed to identify the effects of sex-role stereotyping on children. The survey was conducted with 2,000 Colorado students in 3rd through 12th grades from school districts ranging from metropolitan to rural areas. Students were asked to respond to this question:

"If you woke up tomorrow and discovered that you were a (boy) (girl), how would your life be different?"

Students' grades are identified when this information is specified in the report.

1. Career Choice

Boys' comments

"If I were a girl, I would be expected by some to get married rather than pursue a career." *10th grade.*

"I wouldn't be able to keep my job as a carpenter." *12th grade.*

"I would refuse to work as a secretary or something stupid like that." *11th grade.*

"It would be harder to get a job and I probably would be paid less." *11th grade.*

Girls' comments

"My goal as a girl is to be nothing." *4th grade.*

"I could run for President." *10th grade.*

"I want to be a nurse, but if I were male I would probably want to be an architect." *4th grade.*

"If I were a boy I would be treated better. I would get paid more and be able to do more things." *4th grade.*

2. Appearance

Boys' comments

"I would use a lot of makeup and look good and beautiful to everyone, knowing that few people would care for my personality. The majority of people would like to have me just like a sexual object." *12th grade.*

"I would be treated like a sex symbol." *11th grade.*

"I couldn't be a slob anymore. I'd have to smell pretty." *8th grade.*

Girls' comments

"If I were a boy I wouldn't have to be neat." *4th grade.*

"I wouldn't have to worry about how I look." *6th grade.*

"If I woke up tomorrow and I was a boy, I would go back to bed since it would not take very long to get ready for school." *10th grade.*

3. Violence

Boys' comments

"If I were gorgeous I would be jeered at and hear plenty of comments." *12th grade.*

"I'd have to know how to handle drunk guys and rapists." *8th grade.*

"I would always carry a gun around for protection." *4th grade.*

"I wouldn't have to worry about being scared to fight."

Girls' comments

"I wouldn't have to put up with leers while walking down the street." *11th grade.*

"I wouldn't have to worry about being raped."

"I could beat up people." *6th grade.*

"If I were a boy, I'd *kill* my art teacher instead of arguing with him as I do now." *8th grade.*

4. Peer Relations and Activities

Boys' comments

"No one would make fun of me because I'm afraid of frogs."

"Instead of wrestling with my friends, I'd be sitting around and discussing the daily gossip."

"I couldn't play football or basketball."

"I would like pink and I would like ribbons and pigtails. I would always chase boys. I would like girl teachers better, and would not collect rocks any more. I would be quieter and I'd take a bath in perfume mixed with bubble bath. I wouldn't like being a girl." *3rd grade.*

Girls' comments

"I'd have to put down all the girls." *6th grade.*

"I wouldn't treat chicks like most guys treat me, because I know how it feels."

"I could play baseball or go hunting without being hassled."

5. Personal Behavior and Self-Concept

Boys' comments

"I wouldn't have to worry about responsibilities."

"I would become less outgoing and more polite. I may become shy and be looked upon as a fragile glass doll."

"I would have to hate snakes. Everything would be miserable."

"I'd have to be nicer and say proper things."

"I'd have to be more quiet, more reserved, and wait for others to talk to me." *10th grade.*

Girls' comments

"I'd have to be rowdy, smart alecky, noisy, macho, and say disgusting things."

"I would be more active and show off more."

"I would have to stay calm and cool whenever something happened." *10th grade.*

"I would not be allowed to express my true feelings." *11th grade.*

"I think I would be more outspoken and confident, but I really don't know why." *10th grade.*

6. Freedom and Restrictions

Boys' comments

"I'd have to come in much earlier."

"I couldn't go out as much."

Girls' comments

"I could stay out later."

"There would be fewer rules."

"I'd have more independence."

"I'd be trusted more when driving."

7. Home Life

Boys' comments

"If I were a girl, I would not be able to help my dad fix the car and the truck and his two motorcycles." *6th grade.*

"I'd be the one who has the kid." *8th grade.*

Girls' comments

"Life on the home front would be a lot easier. I know that for a fact since I've got a brother." *4th grade.*

"I would not have to put up with the kids." *6th grade.*

8. School Life

Boys' comments

"I would not want to take all of the math and science courses that I am taking now. I would mostly take art, food, and clothing classes." *10th grade.*

"If I were a (female) athlete, I'd expect fewer people to come to the event." *12th grade.*

"When you're a girl, you cheer sports instead of joining them." *4th grade.*

Girls' comments

"If I was a boy I'd drop my typing class and start taking really hard classes, since my Dad would let me go to college and he won't now." *11th grade.*

"If I were a boy I'd get called on more to answer questions."

"I might be expected to be more intelligent than the girls." *9th grade.*

"I would probably act different toward my teachers, being less cutesy and vulnerable. Boys have to make it on their own." *11th grade.*

"I would use the weight room without feeling funny."

9. Value Judgements

Boys' comments

"Girls can't do anything fun. They don't know how to do anything except play dolls." *4th grade.*

"If I were a girl, I would have to wear make-up, cook, be a mother, and yukky stuff like that." *6th grade.*

"If I were a girl I'd be stupid and weak as a string." *6th grade.*

"If I were a girl, I would want to be a boy." *4th grade.*

"If I were a girl, I'd kill myself."

Girls' comments

"I probably wouldn't get nervous when I talk." *6th grade.*

"I could do stuff better than I do now." *3rd grade.*

"People would take my decisions and beliefs more seriously." *11th grade.*

"I would get married and probably want to have children. I would surely want a son more than a girl." *3rd grade.*

"If I were a boy my father would be closer, because I'd be the son he always wanted." *6th grade.*

"If I were a boy, my Daddy might have loved me." *3rd grade.* □

ALICE I. BAUMGARTNER is an independent organizational consultant and chair of the Advisory Board for the Center for Women's Resources at the University of Colorado at Denver. She is the single parent of two daughters.

SAMPLE ATTITUDE SURVEY QUESTIONS

Adapt questionnaires for math, science, or technology.

For Secondary Students

How important is it to you that you do well in MST? (1 = *very important*, 2 = *important*, 3 = *a little important*, 4 = *not at all important*)

Please rank these subjects according to how important they are. (1 = *most important*, 2 = *important*, 3 = *a little important*, 4 = *not at all imporant*)

_____ Language Arts
_____ Math
_____ Foreign Language
_____ Science
_____ Physical Education
_____ Social Studies
_____ Computer class
_____ Art class

In comparison with other subjects, how much do you like learning MST? (1 = *much more*, 2 = *about the same*, 3 = *much less*)

How much MST do you think most women use in their careers? (1 = *a lot*, 2 = *some*, 3 = *very little*, 4 = *none*)

How much MST do you think most men use in their careers? (1 = *a lot*, 2 = *some*, 3 = *very little*, 4 = *none*)

How much effort does your MST class require compared to your other subjects? (1 = *much less hard*, 2 = *not as hard*, 3 = *about as hard*, 4 = *harder*, 5 = *much harder*)

Compared with your other subjects, how good are you at MST? (1 = *much better*, 2 = *better*, 3 = *about the same*, 4 = *somewhat worse*, 5 = *much worse*)

In comparison with Language Arts, how good are you at MST? (1 = *much better*, 2 = *better*, 3 = *about the same*, 4 = *somewhat worse*, 5 = *much worse*)

Below are possible explanations for when you get a good grade in MST. Please rate them according to how much you agree with each one. (1 = *strongly agree*, 2 = *agree*, 3 = *somewhat agree*, 4 = *somewhat disagree*, 5 = *disagree*, 6 = *strongly disagree*)

___ I am smart in MST.
___ My teacher helped me learn MST.
___ My parents helped me learn MST.
___ I like MST.
___ I worked hard in MST.
___ MST is easy.
___ I used good study skills.

Below are possible explanations for when you get a bad grade in MST. Please rate them according to how much you agree with each one. (1 = *strongly agree*, 2 = *agree*, 3 = *somewhat agree*, 4 = *somewhat disagree*, 5 = *disagree*, 6 = *strongly disagree*)

___ I am not smart in MST.
___ My teacher did not help me learn MST.
___ My parents did not help me learn MST.
___ I don't like MST.
___ I didn't work hard in MST.
___ MST is hard.
___ I didn't use good study skills.

I think I could be successful as a [choose one: mathematician, scientist, engineer, computer specialist]. (1 = *strongly agree*, 2 = *agree*, 3 = *somewhat agree*, 4 = *somewhat disagree*, 5 = *disagree*, 6 = *strongly disagree*)

Adapted from Riesz, Elizabeth D. et al. (1994). "Gender differences in high school students' attitudes toward science: research and intervention" in the Journal of Women and Minorities in Science and Engineering, vol. 1, no. 4, pp. 273-289.

For Elementary Students

What school subject do you like the most? ______

What school subject do you like the least? ______

Do you think your dad is good at MST? ___ Yes ___ No ___ I don't know

How much does your dad talk about MST at home or at work? ___ A lot ___ A little ___ Never ___ I don't know

Do you think your mom is good at MST? ___ Yes ___ No ___ I don't know

How much does your mom talk about MST at home or at work? ___ A lot ___ A little ___ Never ___ I don't know

Who do you think is better at MST? ___ Girls ___ Boys ___ Both the same

Do you get nervous when you have to take a test? ___ Not at all ___ A little ___ More than a little ___ Very nervous

Do you get nervous when you have to take a MSTtest? ___ Not at all ___ A little ___ More than a little ___ Very nervous

How good are you at MST? ___ Much better than most other kids ___ About the same as most other kids ___ Not as good as most other kids ___ Not good at all

How much do you think you will use MST when you grow up? ___ Not at all ___ A little ___ A lot

Do you think knowing MST will help you find a job when you grow up? ___ Definitely ___ I think so ___ I don't think so ___ No

Do your friends think you are good at MST? ___ Definitely ___ I think so ___ I don't think so ___ No

Do you think your teacher likes to teach MST? ___ Not at all ___ A little ___ Pretty much ___ Very much

Do you think MST is important? ___ Yes ___ No

Adapted from Cupillari, Antonella, Hostetler, Robert T. and Tauber, Robert T. (1992). "Attitudes toward mathematics: male/female differences in three grade levels" in New York State Mathematics Teachers Journal, vol. 42, no. 3, pp. 165-172.

MARIA'S BROTHER BOBBY

Maria's brother Bobby was first in his class in advanced placement chemistry. Maria, who is two years younger, demonstrates excellent science skills as well. The science department chair is concerned, however, when Maria does not choose any advanced placement or honors science courses as a high school senior, even though she is well qualified.

When asked, Maria explains that she's interested in the AP course but her parents don't think she should take it. The chair asks the parents to attend a conference. Maria's parents explain to the chairperson that they do not think it is necessary for Maria to take advanced placement science. They fear that it will be too difficult for her and lower her grade point average.

If you were the department chairperson, what would you do to help change Maria's parents' attitudes toward her participation in the chemistry course?

FEDERAL ANTI-DISCRIMINATION LAWS PERTAINING TO SCHOOLS *

First Amendment to the United States Constitution

"Congress shall make no law respecting an establishment of religion, or prohibiting the free exercise thereof, or abridge the freedom of speech ..."

Fourteenth Amendment to the United States Constitution

"No State shall make or enforce any law which shall abridge the privileges or immunities of citizens of the United States; nor shall any State deprive any person of life, liberty, or property without due process of law; nor deny to any person within its jurisdiction the equal protection of the laws."

Equal Pay Act, as amended in 1972

Prohibits discrimination on the basis of sex in wages and fringe benefits by any employer in the United States. The Act provides that a man and a woman working for the same employer under similar conditions in jobs requiring substantially equivalent skills, effort and responsibility must be paid equally even when job titles and assignments are not identical. Employers are required to maintain specified records relevant to the determination of possible violation of the law. Enforced by the U.S. Department of Labor.

Title VI of the Civil Rights Act of 1964

Prohibits discrimination against students on the basis of race, color, or national origin in any school receiving federal assistance. Title VI covers student admissions, access to courses and programs, and student policies and their application. The provision of bilingual instruction or some other method of comprehensible education for students of limited English-speaking ability is also required. Enforced by the Office for Civil Rights, U.S. Department of Education.

Title VII of the Civil Rights Act of 1964
As Amended by the Equal Employment Opportunity Act of 1972

Prohibits discrimination against employees on the basis of sex, race, color, national origin, or religion, by any employer in the United States who employs 15 or more people. This also includes employment agencies, labor unions, and state and local governments. Title VII prohibits discriminatory practices in all terms and conditions of employment, including: recruitment, selection, assignment, transfer, layoff, discharge, and recall; opportunities for promotion; in-service training or development opportunities; wages and salaries; time and pay for sick leave, vacation and overtime; medical, hospital, life and accident insurance; retirement plans and benefits; and other staff benefits. Enforced by the Equal Employment Opportunity Commission, U.S. Department of Labor.

---

* These materials are based on the *Manual for Title IX Coordinators: Sex Equity in New York State Schools*, Division of Civil Rights and Intercultural Relations, 1989, and a subsequent update. Reviewed for accuracy by the ACLU, June 1994.

Pregnancy Discrimination Act of 1978,
An Amendment to Title VII of the Civil Rights Act of 1964

Makes clear that discrimination on the basis of pregnancy, childbirth, or related medical conditions constitute unlawful sex discrimination under Title VII. Enforced by the Equal Employment Opportunity Commission, U.S. Department of Labor.

Executive Order 11246 of 1968
As Amended by Executive Order 11375

Prohibits discrimination against employees on the basis of race, color, religion, sex, or national origin in all schools with federal contracts or subcontracts of $10,000 or more. These orders cover all areas of employment including: hiring, discharge, promotion, wages, benefits, and training. Institutions or agencies with Federal contracts of $50,000 or more and 50 or more employees are also required to develop written affirmative action plans with numerical goals and timetables to cover all employees. Enforced by the Office for Civil Rights, U.S. Department of Education.

Title IX of the Education Amendments of 1972
As Amended by the Civil Rights Restoration Act of 1988

Prohibits discrimination on the basis of sex against any student or employee in all programs of a school district receiving federal financial assistance. Title IX prohibits sex discrimination in such areas as:

- Admission to public undergraduate, graduate, professional, and vocational schools
- Access to courses and programs
- Counseling and guidance practices, tests, and materials
- Physical education and athletics
- Vocational education programs
- Student rules and policies
- Treatment of married and/or pregnant students
- Financial assistance and awards
- Student housing
- Extracurricular activities
- Employment

Agencies and institutions must develop grievance procedures for handling local complaints, appoint a Title IX coordinator, conduct an institutional self-assessment of compliance, and adopt and provide public notification of a policy of nondiscrimination and compliance with Title IX.

In Franklin vs. Gwinnett (1992), the Supreme Court held that those who prove they have experienced sex discrimination, including sexual harassment, can sue their institution for monetary damages under Title IX. There is no cap on the dollar amount of the damages that can be assessed.

Title IX is enforced by the Office for Civil Rights, U.S. Department of Education.

The Rehabilitation Act of 1973

Promotes and expands the opportunities available to individuals with handicapping conditions. Section 502 requires complete accessibility in all buildings constructed after 1968 that were financed with federal funds. Section 503 requires federal contracts valued over $2,500 to include affirmative action and nondiscrimination clauses. Section 504 and its regulations provide equal educational opportunity for "otherwise qualified handicapped individuals" in all educational programs. Equal Educational Opportunity requires program accessibility, identification of needs for educational assistance, and provision for financial assistance to allow students an opportunity to succeed in the school. Enforced by the Office for Civil Rights, U.S. Department of Education.

The Education of All Handicapped Children Act of 1976

Provides federal financial assistance to schools in educating young people between 6 and 21 years of age who have been properly identified and evaluated to be in the targeted categories of: special education, hearing or speech impaired, visually or orthopedically handicapped, emotionally disturbed, or specific learning disabilities. The law requires that students with a handicap must be educated in the "most integrated setting" appropriate. The school must provide each student with an individual education program. Schools are required to search for students with handicaps rather than wait for students to identify themselves to the school. The local education agency, in most cases the school district, is financially responsible for providing all necessary programming for the handicapped student. Agencies must establish a Committee on the Handicapped with parent representation to approve of the educational placement of all handicapped students. A grievance procedure must be established to resolve disputes. Enforced by the Office for Civil Rights, U.S. Department of Education.

Age Discrimination in Employment Act (as Amended) of 1978

Prohibits employers, employment agencies, and labor organizations with 20 or more employees from basing hiring decisions on a person's age when the person's age is between 40 and 70 unless an age limit is a necessary qualification for job performance. Enforced by the U.S. Department of Labor and the Office for Civil Rights, U.S. Department of Education.

The Carl D. Perkins Vocational and Applied Technology Education Act of 1991

Provides funds for vocational education. It places emphasis on making vocational education "... accessible to all persons, including handicapped and disadvantaged persons, single parents and homemakers, ... persons participating in programs designed to eliminate sex bias and stereotyping in vocational education." Among its provisions, the Act seeks to reduce the limiting effects of sex-role stereotyping on occupations, job skills, levels of competency, and careers. It authorizes a state-level sex equity coordinator to administer this portion of the Act. Enforced by the Office for Civil Rights, U.S. Department of Education.

The Americans with Disabilities Act of 1990

Employers with 15 or more employees must offer people with disabilities equal employment opportunities, enabling them by means of reasonable accommodations to attain the same level of performance as employees without disabilities and to enjoy equal benefits and privileges. A disability is "an impairment, physical or mental, that substantially limits one or more major life activities."

The Civil Rights Act of 1991

This act amends and goes beyond Title VII of the Civil Rights Act of 1964 in providing remedies for intentional discrimination and harassment in the workplace on the grounds of sex, race, national origin, disability, and religion.

The Family and Medical Leave Act of 1993

All employers with 50 or more employees to provide up to 12 weeks of unpaid leave per year to eligible employees who want the leave because of the birth of a child or placement of an adopted or foster-care child; to take care of a child, spouse, or parent who has a serious health condition, or for the employee's own serious illness. While on leave, employees are entitled to continued health benefits. Following leave, employers must return employees to the same or an equivalent position.

The School-to-Work Opportunities Act of 1994

Establishes a national framework for school-to-work transition in terms of vocational/occupational training and emphasizes the integration of existing Federal programs such as vocational education and JTPA. State plans and therefore schools are required to 1) provide training of personnel and counseling of women leading to nontraditional job training for women, and 2) specify the methods they will use to ensure that programs funded under this Act are free from racial and sexual harassment.

State with Constitutional Equal Rights Laws

By 1994, sixteen states passed their own versions of the Equal Rights Amendment (which did not pass nationally). They are:

Alaska	New Hampshire
Colorado	New Mexico
Connecticut	Pennsylvania
Hawaii	Texas
Illinois	Utah
Maryland	Virginia
Massachusetts	Washington
Montana	Wyoming

States With Gender Equity in Education Laws

By 1994, fourteen states passed laws that extend Title IX. They are:

Alaska	Minnesota
California	Nebraska
Florida	New Jersey
Hawaii	Oregon
Iowa	Rhode Island
Maine	Washington
Massachusetts	Wisconsin

QUESTIONS FOR TEACHERS WHO WANT TO ENHANCE STUDENT LEARNING

DOROTHY BUERK, ITHACA COLLEGE

1. What are my female students thinking and believing about mathematics? If I don't know, how can I begin to find out?

2. What do I believe, feel, and value about mathematics?

3. Which of my beliefs, feelings, and values would I most like my female students to embrace?

4. How can I help my female students see and experience mathematics in this new way?

5. What resistance do I expect from my female students? What can I do to learn to accept my female students' resistance to change? What can I do to lessen their resistance?

6. What will hinder me from changing my female students' experience with mathematics? (Possibly local or state curriculum guidelines and demands, or the expectations of students, administrators, and other faculty members, etc.) Which of these things can I control? What can I do about them? Which of these are beyond my control? How will they impede my progress?

7. How can I get started? What is most important to me as a first step? What pace should I set? How much can I do and how quickly?

8. What support system do I need to implement my plan? How can I develop that support network for myself?

Reprinted with permission from *Math and Science for Girls*, proceedings of a symposium sponsored by the National Coalition of Girls' Schools. Concord, MA: National Coalition of Girls' Schools, p. 100.

VIGNETTE: AS A HIGH SCHOOL TEACHER

As a high school teacher you regularly teach Basic Mathematics and Advanced Mathematics. One of your best students in the Basic course is Aisha, and you have been encouraging her to sign up for Advanced Mathematics next year.

Today she stays after class to talk to you. "I've been thinking about what you said about taking Advanced next year," she says. "I've talked to a lot of my friends about it. They all say that only boys take it, or maybe boys and a dorky girl. I went to the math club after school. There were only boys there and they acted like jerks. I didn't want to be there so I never went back. I'd hate being the only girl in the class, so I'm not going to take it."

What, as Aisha's teacher, would you say to her? What would you do?

TECHNOLOGY VIGNETTES

A. There is an after-school computer club at your school, and you've been encouraging your female students to join. Several have taken you up on your suggestion, but have dropped out rather quickly. When you ask them why, they tell you that the other club members were mostly boys, that the girls weren't very much interested in what the boys were doing, and in any case were not invited to participate. In addition, the room was plastered with images of males wielding technology, including posters from male-oriented movies and sorcerers and wizards from a variety of battle-oriented computer games.

B. You have two computers in your elementary classroom that are available for students to use as they like. Invariably the boys beat the girls to the machines and then monopolize them for long periods of time, telling the girls that computers are for boys. The girls find other activities that interest them.

C. Several of your female students who have shown an interest in computers and technology have told you that their families are discouraging them from continuing their studies. Their families, they say, are concerned that technology is not "women's work," and that they will not be happy or accepted if they pursue careers in the field.

D. You have encouraged several of your female students to enroll in the advanced programming course at your high school. One of them told you recently that her student advisor counseled her against it, saying that the class was very technical and that she would find it difficult to keep up with the other students (currently all boys).

This article is reprinted with permission from AWIS Magazine, January/February 1996 (vol. 25, no. 1), pp. 8-9. Subscriptions are available through the Association for Women in Science, 1200 New York Avenue, NW, 6th Floor, Washington, DC 20005.

Teaching the MAJORITY

by Sue V. Rosser, PhD, University of Florida
Director of the Center for Women's Studies and Gender Research

During the last two decades, women have entered the professions in record numbers. In recent years, women have received degrees in most fields in numbers approaching or exceeding their 51% of the American population. The physical sciences, mathematics, and engineering persist as the professional areas where women have not yet broken the gender barrier. Women constitute 45% of the employed labor force in the United States but only 16% of all employed scientists and engineers.[1] A 1994 report from the National Research Council revealed that women constitute about 12% of the employed scientific and engineering labor force in industry.[2] Of the 1,647 living scientists elected to membership in the National Academy of Sciences, only 70 are women; in the 1992 election, 5 of the 59 honorees were women.

Despite the relatively low percentages of women in most areas of science and engineering, until recently, few programs have directly targeted females. The results of a 1991 study by Matyas and Malcom, which included surveys of the presidents and chancellors of 276 colleges and universities and the directors of nearly 400 recruitment and retention programs, revealed that less than 10% of the programs included in the study were specifically focused on the recruitment and retention of women in science or engineering.[3] This study confirmed similar findings from previous studies that virtually no programs directly target female students or faculty.

A growing body of research documents the need to change the way science is taught in order to appeal to women. Women's studies scholars have explored the ways in which science as it is currently taught and practiced may reflect a masculine approach to the world that tends to exclude women.[4,5] This critique has been developed most extensively for biology,[6-11] leading scholars to examine curricular content and pedagogy in that discipline.

Based on these investigations, faculty have evolved new approaches to teaching traditional material that is more "female-friendly".[11,12] In two previous volumes in the Athene Series, I explored revisions of biology and health curricular content and syllabi[10] and pedagogical techniques[11] to include women. After reading these books or hearing me speak about them, people frequently request similar information for the physical sciences, mathematics, and engineering.

> **A growing body of research documents the need to change the way science is taught in order to appeal to women.**

This volume is an attempt to fill that request. Knowing that such curricular and pedagogical innovations had to be developed by individuals who teach the subjects, I agreed to edit the volume with chapters contributed by mathematicians, engineers, physicists, chemists, computer scientists, and geologists who teach in colleges and universities throughout the United States. *Teaching the Majority* includes descriptions of changed teaching techniques, course content, syllabi, laboratory exercises, and problem sets demonstrated to attract and retain women. Each chapter is written by a faculty member who has successfully transformed his or her science, mathematics, or engineering course to appeal to women students in particular, while retaining its appeal for male students.

Although it would be desirable to include a course description and syllabus, examples of laboratory exercises and problem sets, and specific pedagogical techniques for each discipline within the physical sciences and engineering, the cutting-edge nature of this volume makes that impossible. A few faculty members in the physical sciences and engineering have only recently begun to consider the impact of gender and the application of the new scholarship on women in their classrooms. It is not surprising that one faculty member working in isolation—not only alone in her or his department and institution but also unsupported by a national professional organization or project—may make a considerable contribution to the field by describing a new pedagogical technique or transformed curricular content of one subsection of a traditional syllabus. The uneven extent of the transformation for the different disciplines is inevitable in this cutting-edge work.

Twenty-five years of women's studies scholarship and experience with curriculum transformation projects have enabled faculty to develop models that chart the phases through which changes occur in a variety of disciplines in diverse institutions.[13-15] Building upon these models for other disciplines, I developed similar phase models and pedagogical techniques for the sciences.[11,12] The phases of curriculum transformation may be visualized as a continuous spiral with overlapping components rather than as discrete stages; many of the pedagogical techniques are appropriate to accompany multiple stages of the curriculum. A six-phase model for curricular and pedagogical transformations provides a framework to explore contributions by the authors of *Teaching the Majority.*

PHASE I: Absence of women from the curriculum is not noted.

PHASE II: Recognition that most scientists are male and that science may reflect a masculine perspective.

Pedagogical techniques to accompany this phase:

1. Undertake fewer experiments likely to have applications of direct benefit to the military and propose more experiments to explore problems of social concern.
2. Include problems that have not been considered worthy of scientific investigation because of the field with which they have been traditionally associated (e.g., home economics and nursing).
3. Undertake the investigation of problems with a more holistic, global scope and use interactive methods to approach them rather than the more reduced and limited scale problems traditionally considered.

PHASE III: Identification of barriers that prevent women from entering science.

Pedagogical techniques to accompany this phase:

1. Expand the kinds of observations beyond those traditionally carried out in scientific research.
2. Increase the numbers of observations and remain longer in the observation stage of the scientific method.
3. Incorporate and validate women's personal experiences as part of class discussions or laboratory exercises.

PHASE IV: Search for women scientists and their unique contributions.

Pedagogical techniques to accompany this phase:

1. Include the names of women scientists who have made important discoveries.
2. Use less competitive models and more interdisciplinary methods to teach science.
3. Discuss the role of scientist as only one facet that must be smoothly integrated with other aspects of students' lives.
4. Put increased effort into strategies such as teaching and communicating with nonscientists to break down barriers between science and the layperson.
5. Discuss the practical uses to which scientific discoveries are put to help students see science in its social context.

PHASE V: Science done by feminists and women.

Pedagogical techniques to accompany this phase:

1. Use precise, gender-neutral language in describing data and presenting theories.
2. Encourage development of theories and hypotheses that are relational, interdependent, and multicausal rather than hierarchical, reductionistic, and dualistic.
3. Use a combination of qualitative and quantitative methods in data collection.
4. Encourage the uncovering of biases such as race, class, sexual orientation, and religious affiliation, as well as gender, which may permeate theories and conclusions drawn from experimental observation.

The physical sciences, mathematics, and engineering persist as the professional areas where women have not yet broken the gender barrier.

PHASE VI: Science redefined and reconstructed to include us all.

The transformed curricula, expanded problem sets and laboratory exercises, and successful pedagogical techniques presented by the authors of *Teaching the Majority* were contributed to aid others in breaking the gender barrier in science and technology. The authors hope to encourage faculty members at colleges, universities, and institutes of technology who teach courses in geology, chemistry, physics, mathematics, computer science, environmental science, and engineering on both the undergraduate and graduate levels to undertake similar initiatives in the courses they teach. Deans, department chairs, and other administrators may use *Teaching the Majority* as a foundation for faculty development workshops designed to integrate the new scholarship on women into the physical sciences, mathematics, and engineering. Many professionals outside of academia, such as managers in industry and government who give on-the-job training to engineers, scientists, and technicians, will find new models and ideas to use.

The authors of *Teaching the Majority* have explored changing curricula and teaching techniques in the hope of producing a different composition in the pool of scientists—scientists who hold a slightly modified theoretical perspective. This perspective may in turn be reflected in further transformation of curricula and teaching techniques to make science more attractive to women. The ultimate end of this upward spiral would be the creation of a community of scientists that proportionately represents the diversity of the population as a whole with regard to gender, race, and class. ❖

REFERENCES

1. National Science Foundation. 1992. *Women and Minorities in Science and Engineering: An Update* (NSF 92-303). Washington, DC: Author.
2. National Research Council. 1994. *Women Scientists and Engineers Employed in Industry: Why So Few?* Washington, DC: Author.
3. Matyas M. and S. Malcom. 1991. *Investing in Human Potential: Science and Engineering at the Crossroads.* Washington, DC: American Association for the Advancement of Science.
4. Harding, S. 1986. *The Science Question in Feminism.* Ithaca: Cornell University Press.
5. Keller, E. F. 1985. *Reflections on Gender and Science.* New Haven: Yale University Press.
6. Birke, L. 1986. *Women, Feminism, and Biology: The Feminist Challenge.* New York: Metheun.
7. Bleier, R. 1984. *Science and Gender: A Critique of Biology and its Theories on Women.* Elmsford, NY: Pergamon Press.
8. Fausto-Sterling, A. 1992. *Myths of Gender.* New York: Basic Books.
9. Hubbard, R. 1990. *The Politics of Women's Biology.* New Brunswick: Rutgers University Press.
10. Rosser, S. V. 1986. *Teaching Science and Health from a Feminist Perspective: A Practical Guide.* Elmsford: Pergamon Press.
11. Rosser, S. V. 1990. *Female-friendly Science.* Elmsford: Pergamon Press.
12. Rosser, S. V. 1993. *The Journal of General Education* 42: 191-220.
13. McIntosh, P. 1984. *Forum for Liberal Education* 6:2-4.
14. Schuster M. and S. Van Dyne. 1984. *Harvard Educational Review* 54:413-428.
15. Tetreault, M. K. 1985. *Journal of Higher Education* 5:368-384.

This article is reprinted in modified form with permission of the author from Teaching the Majority, *Teachers College Press, NY, NY.*

THE FORMS OF BIAS IN INSTRUCTIONAL MATERIALS

INVISIBILITY	Certain groups are underrepresented in curricular materials. The significant omission of women and minority groups implies that they are of less value, importance, and significance in our society.
STEREOTYPING	By assigning traditional and rigid roles or attributes to a group, instructional materials limit the abilities and potential of that group. Stereotyping denies students a knowledge of the diversity, complexity, and variation of any group of individuals. Children who see themselves portrayed only in stereotypic ways may internalize these stereotypes and fail to develop their own unique abilities, interests, and full potential.
IMBALANCE	Some texts perpetuate bias by presenting only one interpretation of an issue, situation, or group of people. This selectivity restricts students' knowledge of the varied perspectives that actually apply, thus ignoring differing viewpoints, reducing complexity to a simplistic level, and distorting reality.
UNREALITY	Some texts present an unrealistic portrayal of our history and our contemporary life experiences. Controversial topics are glossed over and discussion of uncomfortable realities, such as discrimination, bias and prejudice, are avoided. This incomplete or unrealistic coverage denies children the information they need to recognize, understand, and perhaps some day conquer the problems that plague our society.
FRAGMENTATION	By isolating issues related to women and minorities from the main body of the text in sidebars or boxes, instructional materials imply that these issues are less important than and separate from the rest of the topics and the cultural mainstream.
LINGUISTIC BIAS	Curricular materials reflect the biased nature of our language. Masculine terms and pronouns, such as the term forefathers or the generic he, deny the participation of women in our society. Occupations such as fireman are given masculine labels that deny the legitimacy of women working in these fields. Imbalance of word order and lack of parallel terms that refer to females and males are also forms of linguistic bias.
"COSMETIC" BIAS	Textbook layouts can falsely suggest that women and minorities are included equally throughout the text, perhaps by prominently displaying pictures and highlighted sections featuring their experiences and accomplishments, while a careful analysis of the six factors above shows that these groups actually receive little coverage. This is especially an issue with more recent materials.

Adapted by permission from: Sadker, Myra P. and Sadker, David M. (1982). *Sex equity handbook for schools.* New York: Longman Inc., pp. 72-73. Also Sadker, Myra and Sadker, David (1995). *Gender bias in the curriculum.* Washington, DC: National Education Association. p. 11.

VIGNETTE: YOU ARE ABOUT TO START YOUR STUDENT TEACHING

You are about to start your student teaching in Ms. Smith's classroom. You have spent considerable time learning about gender equity in mathematics education in your methods course, and how biased materials can discourage girls from persisting in mathematics.

When you enter the classroom, you can't help but notice the bulletin boards are decorated with pictures of famous male mathematicians. The textbook the children are using contains many gender-stereotyped themes such as girls calculating recipe quantities, boys calculating sports averages. The supplementary print and audiovisual materials used with the class are also gender biased.

You are disturbed to notice that the girls seem to participate less in class discussion. They seem to sit back and let the boys answer questions, and seem less interested overall in the class than the boys.

What would you do about this situation? Why?

REAL-LIFE MATHEMATICS

Find out what mathematics women use at work and at home, and why it is important to them.

Directions:

1. Read over the following list of math skills. Discuss examples that illustrate use of each skill.

   - Angle measurement
   - Area and perimeter
   - Averaging
   - Calculators
   - Coordinate graphing
   - Circles
   - Decimals
   - Estimation
   - Exponents and scientific notation
   - Formulas
   - Fractions
   - Geometric concepts
   - Linear equations
   - Percentages
   - Probability
   - Pythagorean Theorem
   - Ratio and proportion
   - Statistical graphs
   - Trigonometric functions
   - Vectors

2. Look at sample directions for the activities that are considered traditional "women's work."

3. Complete the Useful Math Data Sheet on the next page.

USEFUL MATH DATA SHEET

TASK	MATHEMATICS USED	OTHER USEFUL SKILLS	COMMENTS

NCTM STANDARDS

The National Council of Teachers of Mathematics' Professional Teaching Standards states the following:

"Teachers also need to understand the importance of context as it relates to students' interest and experiences. Instruction should incorporate real-world contexts and children's experiences, and, when possible, should use children's language, viewpoints, and culture..."

In the space below, list what you would do as a teacher to identify your female students' interests and experiences.

GUIDELINES FOR A CHILDREN'S SURVEY

a. Invite the students to break into small working groups. Each group chooses a topic relevant to themselves and their peers on which to base a survey. Topics can be anything of interest chosen by the group.

b. Each group designs and administers its survey and decides how the computer can be used in the process. Students can use word processors or desktop publishing programs to prepare survey forms, and spreadsheets, graphics programs, and graphing programs to represent their findings. Advanced students can use a statistical analysis program to analyze their data.

c. Invite students to share their findings with one another, either by preparing copies of their results for everyone or by using them in a bulletin board.

Guidelines for Nonsexist Use of Language in NCTE Publications (Revised, 1985)

Introduction

During the 1971 Annual Convention of the National Council of Teachers of English in Las Vegas, Nevada, the Executive Committee and the Board of Directors approved the formation of an NCTE Committee on the Role and Image of Women in the Council and the Profession. As the result of a resolution passed by the members of NCTE at the 1974 Annual Convention, one of the committee's responsibilities was to assist in setting guidelines for nonsexist* use of language in NCTE publications.

Suggestions were elicited from editors of Council journals and from professional staff members at NCTE, as well as from members of the Women's Committee. Copies of the guidelines also went to all members of the Board of Directors. At the 1975 Annual Convention, the Board of Directors adopted a formal policy statement that read in part: "The National Council of Teachers of English should encourage the use of nonsexist language, particularly through its publications and periodicals."

Ten years have passed since these guidelines were created, and although language usage has begun to change, the importance of the guidelines has not diminished. Because language plays a central role in the way human beings think and behave, we still need to promote language that opens rather than closes possibilities for women and men. Whether teaching in the classroom, assigning texts, determining curriculum, serving on national committees, or writing in professional publications, NCTE members directly and indirectly influence thought and behavior.

As an educational publisher, NCTE is not alone in its concern for fair treatment of men and women. The role of education is to make choices available, not to limit opportunities. Censorship removes possibilities; these guidelines extend what is available by offering alternatives to traditional usages and to editorial choices that restrict meaning.

Language

This section deals primarily with word choice. Many of the examples are matters of vocabulary; a few are matters of grammatical choice. The vocabulary items are relatively easy to deal with, since the English lexicon has a history of rapid change. Grammar is a more difficult area, and we have chosen to use alternatives that already exist in the language rather than to invent new constructions. In both cases, recommended alternatives have been determined by what is graceful and unobtrusive. The purpose of these changes is to suggest alternative styles.

*Although *nonsexist* is the word traditionally used to describe such language, other terms have come into common use, namely, *gender-neutral, sex-fair, gender-free.*

Generic "Man"

1. Since the word *man* has come to refer almost exclusively to adult males, it is sometimes difficult to recognize its generic meaning.

Problems	*Alternatives*
mankind	humanity, human beings, people*
man's achievements	human achievements
the best man for the job	the best person for the job
the common man	the average person, ordinary people
cavemen	cave dwellers, prehistoric people

2. Sometimes the combining form *-woman* is used alongside *-man* in occupational terms and job titles, but we prefer using the same titles for men and women when naming jobs that could be held by both. Note, too, that using the same forms for men and women is a way to avoid using the combining form *-person* as a substitute for *-woman* only.

Problems	*Alternatives*
chairman/chairwoman	chair, coordinator (of a committee or department), moderator (of a meeting), presiding officer, head, chairperson
businessman/businesswoman	business executive, manager
congressman/congresswoman	congressional representative
policeman/policewoman	police officer
salesman/saleswoman	sales clerk, sales representative, salesperson
fireman	fire fighter
mailman	letter carrier

Generic "He" and "His"

Because there is no one pronoun in English that can be effectively substituted for *he* or *his*, we offer several alternatives. The form *he or she* has been the NCTE house style over the last ten years, on the premise that it is less distracting then *she or he* or *he/she*. There are other choices, however. The one you make will depend on what you are writing.

1. Sometimes it is possible to drop the possessive form *his* altogether or to substitute an article.

Problems	*Alternatives*
The average student is worried about his grades.	The average student is worried about grades.
When the student hands in his paper, read it immediately.	When the student hands in the paper, read it immediately.

2. Often, it makes sense to use the plural instead of the singular.

Problems	*Alternatives*
Give the student his grade right away.	Give the students their grades right away.
Ask the student to hand in his work as soon as he is finished.	Ask students to hand in their work as soon as they are finished.

*A one-word substitution for *mankind* isn't always possible, especially in set phrases like *the story of mankind.* Sometimes recasting the sentence altogether may be the best solution.

3. The first or second person can sometimes be substituted for the third person.

Problems	*Alternatives*
As a teacher, he is faced daily with the problem of paperwork.	As teachers, we are faced daily with the problem of paperwork.
When a teacher asks his students for an evaluation, he is putting himself on the spot.	When you ask your students for an evaluation, you are putting yourself on the spot.

4. In some situations, the pronoun *one (one's)* can be substituted for *he (his)*, but it should be used sparingly. Notice that the use of *one*—like the use of *we* or *you*—changes the tone of what you are writing.

Problem	*Alternative*
He might well wonder what his response should be.	One might well wonder what one's response should be.

5. A sentence with *he* or *his* can sometimes be recast in the passive voice or another impersonal construction.

Problems	*Alternatives*
Each student should hand in his paper promptly.	Papers should be handed in promptly.
He found such an idea intolerable.	Such an idea was intolerable.

6. When the subject is an indefinite pronoun, the plural form *their* can occasionally be used with it, especially when the referent for the pronoun is clearly understood to be plural.

Problem	*Alternative*
When everyone contributes his own ideas, the discussion will be a success.	When everyone contributes their own ideas, the discussion will be a success.

But since this usage is transitional, it is usually better to recast the sentence and avoid the indefinite pronoun.

Problem	*Alternative*
When everyone contributes his own ideas, the discussion will be a success.	When all the students contribute their own ideas, the discussion will be a success.

7. Finally, sparing use can be made of *he or she* and *his or her*. It is best to restrict this choice to contexts in which the pronouns are not repeated.

Problems	*Alternatives*
Each student will do better if he has a voice in the decision.	Each student will do better if he or she has a voice in the decision.
Each student can select his own topic.	Each student can select his or her own topic.

Sex-Role Stereotyping

Word choices sometimes reflect unfortunate and unconscious assumptions about sex roles—for example, that farmers are always men and elementary school teachers are always women; that men are valued for their accomplishments and women for their physical attributes; or that men are strong and brave while women are weak and timid. We need to examine the assumptions inherent in certain stock phrases and choose nonstereotyped alternatives.

1. Identify men and women in the same way. Diminutive or special forms to name women are usually unnecessary. In most cases, generic terms such as *doctor* or *actor* should be assumed to include both men and women. Only occasionally are alternate forms needed, and in these cases, the alternate form replaces both the masculine and the feminine titles.

Problems	*Alternatives*
stewardess	flight attendant (for both *steward* and *stewardess*)
authoress	author
waitress	server, food server
poetess	poet
coed	student
lady lawyer	lawyer . . . she
male nurse	nurse . . . he

2. Do not represent women as occupying only certain jobs or roles and men as occupying only certain others.

Problems	*Alternatives*
the kindergarten teacher . . . she	*occasionally use* the kindergarten teacher . . . he *or* kindergarten teachers . . . they
the principal . . . he	*occasionally use* the principal . . . she *or* principals . . . they
Have your mother send a snack for the party.	Have a parent send a snack for the party. *occasionally use* Have your father . . . *or* Have your parents. . . .
NCTE conventiongoers and their wives are invited.	NCTE conventiongoers and their spouses are invited.
Writers become so involved in their work that they neglect their wives and children.	Writers become so involved in their work that they neglect their families.

3. Treat men and women in a parallel manner.

Problems	*Alternatives*
The class interviewed Chief Justice Burger and Mrs. O'Connor.	The class interviewed Warren Burger and Sandra O'Connor. *or* . . . Mr. Burger and Ms. O'Connor. *or* . . . Chief Justice Burger and Justice O'Connor.
The reading list included Proust, Joyce, Gide, and Virginia Woolf.	The reading list included Proust, Joyce, Gide, and Woolf. *or* . . . Marcel Proust, James Joyce, André Gide, and Virginia Woolf.
Both Bill Smith, a straight-A sophomore, and Kathy Ryan,	Both sophomore Bill Smith, a straight-A student, and junior

a pert junior, won writing awards.	Kathy Ryan, editor of the school paper, won writing awards.

4. Seek alternatives to language that patronizes or trivializes women, as well as to language that reinforces stereotyped images of both women and men.

Problems	*Alternatives*
The president of the company hired a gal Friday.	The president of the company hired an assistant.
I'll have my girl do it.	I'll ask my secretary to do it.
Stella is a career woman.	Stella is a professional. *or* Stella is a doctor (architect, etc.).
The ladies on the committee all supported the bill.	The women on the committee all supported the bill.
Pam had lunch with the girls from the office.	Pam had lunch with the women from the office.
This is a man-sized job.	This is a big (huge, enormous) job.
That's just an old wives' tale.	That's just a superstition (superstitious story).
Don't be such an old lady.	Don't be so fussy.

Sexist Language in a Direct Quotation

Quotations cannot be altered, but there are other ways of dealing with this problem.

1. Avoid the quotation altogether if it is not really necessary.
2. Paraphrase the quotation, giving the original author credit for the idea.
3. If the quotation is fairly short, recast it as an indirect quotation, substituting nonsexist words as necessary.

Problem	*Alternative*
Among the questions asked by the school representatives was the following: "Considering the ideal college graduate, what degree of knowledge would you prefer him to have in each of the curricular areas?"	Among the questions asked by the school representatives was one about what degree of knowledge the ideal college graduate should have in each of the curricular areas.

Sample Revised Passage

Substantial revisions or deletions are sometimes necessary when problems overlap or when stereotyped assumptions about men and women so pervade a passage that simple replacement of words is inadequate.

Problem	*Alternative*
Each student who entered the classroom to find himself at the mercy of an elitist, Vassar-trained Miss Fidditch could tell right away that the semester would be a trial. The trend in composition pedagogy toward student-centered essays and away from hours of drill on grammatical correctness has meant, at least for him, that he can finally learn to write. But Macrorie, Elbow, and Janet Emig could drive the exasperated teacher of a cute and perky cheerleader type to embrace the impersonal truth of *whom* as direct object rather than fight his way against the undertow of a gush of personal experience. As Somerset Maugham remarked, "Good prose should resemble the conversation of a well-bred man," and both Miss Fidditch and the bearded guru who wants to "get inside your head" must realize it.	The trend in composition pedagogy toward student-centered essays, represented by such writers as Ken Macrorie, Peter Elbow, and Janet Emig, has meant that some students are finally learning to write. Yet the movement away from hours of drill on grammatical correctness has brought with it a new problem: in the hands of the inexperienced teacher, student essays can remain little more than unedited piles of personal experiences and emotions.

Representation of Men and Women

Important as language is, striving for nonsexist usage is to little purpose if the underlying assumptions about men and women continue to restrict them to traditional roles. If women never enter an author's world, for example, it little avails a writer or editor to refer scrupulously to students as "they" and prehistoric people as "cave dwellers." Thus, teachers and other professionals must be alert to the possible sexist implications of the content as well as the language of educational materials.

It has been enheartening to note that in the last ten years, trade publishers, textbook publishers, and publishers of reference works have become acutely aware of sexist language, thus largely alleviating the problem of discriminatory reference. Still, vigilance must be exercised.

The following recommendations concerning educational materials are made to correct traditional omissions of women or perpetuations of stereotypes.

Booklists

1. Items for a booklist should be chosen to emphasize the equality of men and women and to show them in nontraditional as well as traditional roles. Many children's favorites and classics may contain sexist elements, but books that are valuable for other reasons should not be excluded. The annotations, however, should be written in nonsexist language.
2. Picture books should be chosen showing males and females actively participating in a variety of situations at home, work, and play.
3. Booklists should be organized by subject headings that do not assume stereotyped male and female interests.

Problems	*Alternatives*
Books for Boys	Arts and Crafts
Books for Girls	Sports
	Travel

Teaching Units

1. The topic and organization of teaching units should be carefully considered to avoid sexist implications. Literature by and about both women and men should be included wherever possible.
2. When materials are chosen that present stereotyped assumptions about men and women, they should be balanced by others that show nontraditional roles and assumptions. *Jemima Puddle-Duck* and *Peter Rabbit* read together, for instance, show foolishness is not a sex-linked characteristic. Vera Brittain's *A Testament of Youth* and Ernest Hemingway's *The Sun Also Rises* present the aftermath of World War I from provocative perspectives. Placing a book in the proper historical context and using discussion questions that reflect an awareness of the sexist elements are good strategies.
3. Activities suggested in teaching units should not be segregated by sex: boys can make costumes and girls can build sets.

Reference Books and Research Materials

Reference books can be implicitly sexist in their titles, organizations, content, and language. Editors of such books should follow the suggestions in this publication to ensure nonsexist language in bibliographies, indexes, style manuals, and teacher's guides. In research works, if both males and females were studied, references to individual subjects should not assume that they are all one sex.

Implementation of Guidelines

These guidelines for nonsexist language are suggestions for teachers, writers, and contributors to NCTE publications. For the editors of NCTE publications, however, they are a statement of editorial policy.

Traditionally, editors have set the style for their publications–deciding, for example, whether there should be a comma before the conjunction in a series or whether the first item in a list after a colon should begin with a capital letter. Style decisions have sometimes been made in response to public pressure. Writing *Negro* with a capital letter instead of a lowercase letter and, later, using *Black* instead of *Negro* were both style decisions of this sort for many publishing houses, newspapers, and magazines.

It is an editor's job to rewrite whenever necessary to eliminate awkward language, inconsistency, or inaccuracy. If a job title is inaccurately identified in an article as Director of Public Instruction but the title is actually Supervisor of Public Instruction, the editor changes the wording as a matter of course and without asking the author's approval. If the subject matter or tone of an article is totally inappropriate for the particular publication, it would also be the editor's prerogative to return the manuscript to the author. In the case of language inconsistent with the guidelines, it is the editor's duty to question the author's use of a particular term; on the other hand, the author has the right to insist on its use, but a footnote will be provided to reflect such insistence.

The choices suggested in these guidelines are intended as additions to the style sheets and manuals already in use.

References

Authors and editors who would like to see further suggestions for creating a graceful, nondiscriminatory writing style should refer to these publications. (Note that many of the publishers' guidelines are in the process of being revised.)

American Psychological Association Task Force on Issues of Sexual Bias in Graduate Education. "Guidelines for Nonsexist Use of Language." *American Psychologist* 30 (June 1975): 682–84.

Editorial and Art Content Criteria for Treatment of Minorities and Women. Lexington: Ginn and Company. (Available from the publisher, 191 Spring Street, Lexington, MA 02173.)

Fairness in Educational Materials: Exploring the Issues. Chicago: Science Research Associates, Inc. (Available from the publisher, 155 North Wacker Drive, Chicago, IL 60606.)

Guidelines for Bias-Free Publishing. New York: McGraw-Hill Book Company. (Available from the publisher's distribution center, Princeton Road, Hightstown, NJ 08520.)

Guidelines for Creating Positive Sexual and Racial Images in Educational Materials. New York: Macmillan Publishing Company, 1975. (Available in limited quantities from the publisher, 866 Third Avenue, New York, NY 10022.)

Guidelines for Developing Bias-Free Instructional Materials. Morristown: Silver Burdett Company, 1979. (Available from the publisher, 250 James Street, Morristown, NJ 07960.)

Guidelines for the Development of Elementary and Secondary Instructional Materials. New York: Holt, Rinehart and Winston School Department, 1975. (Available from the publisher, 383 Madison Avenue, New York, NY 10017.)

Miller, Casey, and Kate Swift. *The Handbook of Nonsexist Writing: For Writers, Editors and Speakers.* New York: Barnes and Noble Books, 1980. (Available from Harper and Row, 10 East 53rd Street, New York, NY 10022.)

Nilsen, Alleen Pace. "Editing for Sex." *Idaho English Journal* 6 (Spring 1983): 12+.

_____. "Winning the Great *He/She* Battle." *College English* 46 (February 1984): 151.

Statement on Bias-Free Materials. New York: Association of American Publishers. (Available from AAP, One Park Avenue, New York, NY 10016.)

Additional copies of *Guidelines for Nonsexist Use of Language in NCTE Publications* are available from NCTE, 1111 Kenyon Road, Urbana, Illinois 61801. Single copies are available free upon request, and may be copied without permission from NCTE. Please enclose a self-addressed, stamped envelope. Multiple copies are available in groups of 100 at a bulk rate of U.S. $7 per 100, prepaid only. Ask for Stock No. 19719-012.

GENDER-BIASED STUDENT/TEACHER INTERACTIONS

Studies have documented patterns in which teachers ...

QUESTIONS AND ANSWERS

- Call on boys more than girls
- Accept boys' called-out answers more than girls'
- Wait longer for boys' answers than girls'
- Give girls more neutral responses ("Okay") and boys more complex responses, both postive and negative
- Allow boys more talking time than girls
- Frown more during girls' answers than boys'
- Allow themselves to be interrupted more easily when girls are speaking than when boys are
- Look at their watches or a clock more frequently when girls are speaking than when boys are

PRAISE, CRITICISM, AND FEEDBACK

- Praise girls for the form or appearance of their work, and boys for the content of their work
- Tell boys how to solve problems, but solve the problems for girls: learned helplessness
- Discipline boys more than girls even when they misbehave equally
- Give boys more criticism and corrective feedback

PHYSICAL MOVEMENT

- Position their bodies toward boys more than girls
- Circulate more to boys' seats than to girls' seats, or to the boys' "area" of the room when children are allowed to sex-segregate their seating pattern

OTHER

- Permit students to make gender-biased behaviors or comments to each other
- Allow students to self-segregate by sex
- Give girls fewer experiences with science instruments and equipment
- Allow boys to have more time with scarce classroom resources such as computers or science equipment
- Assign different tasks on the basis of gender

Jo Sanders, 1996

TEACHER/STUDENT INTERACTION CODING FORM 1

Behavior observed ______________________________ Date _______________

Teacher ____________________________ Observer _____________________________

Instruction: Make a cross-hatch mark each for each codable incident.

GIRLS	BOYS

Jo Sanders, 1996

TEACHER/STUDENT INTERACTION CODING FORM 2

Date ______________ Teacher ________________________________

Time observation started ___________ Time observation ended ___________

Description of lesson and instructional strategies:

Student name	Gender	Race	Called On	Called Out

Patricia B. Campbell, 1996

TEACHER/STUDENT INTERACTION CODING FORM 3

Date ______________ Teacher ________________________________

Time observation started ___________ Time observation ended __________

Description of lesson and instructional strategies:

Student name	Gender	Race	Called On	Called Out	Disci-plined	Praised for Answer	Praised for Other Things

Patricia B. Campbell, 1996

TEACHER/STUDENT INTERACTION CODING FORM 4

VERBAL INTERACTION CATEGORIES

Instruction: Fill out the form on the next page according to these verbal interaction categories.

PRAISE	ACADEMIC	Teacher rewards and reinforcement given directly for the intellectual quality of academic work. For example: "Good answer," or "You've written a very interesting report."
	NON-ACADEMIC	Teacher rewards and reinforcement that are not directed to the intellectual quality of the work, such as "You're being nice and quiet today."
ACADEMIC CRITICISM	INTELLECTUAL QUALITY	Critical remarks directed at the lack of intellectual quality of work, such as "Perhaps mathematics isn't a good field for you."
	EFFORT	Teacher comments attributing academic failure to lack of effort, such as "You're not trying hard enough."
NON-ACADEMIC CRITICISM	MILD	Negative teacher comments that reprimand violations of conduct, rules, forms, behaviors, and other nonacademic areas, such as "Tom, stay in line."
	HARSH	Negative comments that attract attention because they are louder, longer, and/or stronger than mild criticism, such as "Tom, get back in line. I have had more than enough from you today. Stay in line or suffer the consequences. MOVE."
QUESTIONS	LOW-LEVEL	Teacher questions that require memory on the part of the student, such as "Who was the fifth president of the United States?"
	HIGH-LEVEL	Teacher questions that require intellectual processes and ask the student to use the information, not just memorize it, such as "What would you include in your personal statement on animal rights?"
ACADEMIC INTERVEN-TION	FACILITATE	Teacher behaviors that facilitate learning by providing students with suggestions, hints, and cues to encourage and enable them to complete the assignment themselves, such as "Think of yesterday's problem and try that one again."
	SHORT-CIRCUIT	Comments that prevent or short-circuit student success by taking over the learning process, such as "That's wrong. The answer is 14."

Daniel P. Shepardson and Edward L. Pizzini,
Science and Children, Nov./Dec. 1991

Page 2

TEACHER/STUDENT INTERACTION CODING FORM 4
VERBAL INTERACTION CATEGORIES

VERBAL INTERACTIONS FORM

VERBAL BEHAVIOR		DIRECTED AT BOYS	DIRECTED AT GIRLS
PRAISE	ACADEMIC		
	NON-ACADEMIC		
ACADEMIC CRITICISM	INTELLECTUAL		
	EFFORT		
NON-ACADEMIC CRITICISM	MILD		
	HARSH		
QUESTIONS	LOW-LEVEL		
	HIGH-LEVEL		
ACADEMIC INTERVEN-TION	FACILITATE		
	SHORT-CIRCUIT		

Daniel P. Shepardson and Edward L. Pizzini, 1991

TEACHER/STUDENT INTERACTION CODING FORM 5

GESA (GENDER/ETHNIC EXPECTATIONS AND STUDENT ACHIEVEMENT) OBSERVATION FORM

Interactions: a ________________ b__________________ Class __________

				Observ. 1		Observ. 2		Observ. 3		Total	
Student No.	Sex	First Name	Eth.	a	b	a	b	a	b	a	b
01											
02											
03											
04											
05											
06											
07											
08											
09											
10											
11											
12											
13											
14											
15											
16											
17											
18											
19											
20											
21											
22											
23											
24											
25											

VIGNETTE: MS. MARCUS' FIFTH-GRADE CLASS

Ms. Marcus' fifth-grade class is making presentations to the class on inventors.

She asks who wants to go first. Deirdre raises her hand and gives her presentation about two chemists who became famous cosmetologists: Madame C. J. Walker and Helena Rubinstein. She displays a poster showing some of the chemical processes required to make face cream. Her presentation lasts seven minutes. There are no questions, but everyone applauds and she sits down.

Ronald goes next. He is dressed in a white lab coat and he talks about Charles Drew, who invented the blood bank system. He also has posters with him, and his presentation lasts 15 minutes. Ms. Marcus asks him several probing questions, and then the class asks other questions. Ronald is in front of the class 25 minutes before he sits down.

If you were Ms. Marcus, what would you do about the unequal time children have for their presentations?

SCENARIO: YOU ARE TEACHING YOUR CLASS

You are teaching your class and have just called on Junie to answer a question. She looks confused. Before Junie has a chance to say anything, Mark mutters loudly enough for everyone to hear, "She'll never get it — she's an airhead."

You could:

- Ignore the comment.
- Disagree, saying that Junie isn't stupid at all.
- Agree, saying that after all, Junie is trying hard.
- Reprimand Mark mildly.
- Reprimand Mark harshly.
- Tell Mark to leave your class and return only when he can be civil.
- Hold a class discussion on insulting behavior.

Consider the consequences for Junie, Mark, and the rest of the class of various courses of action open to you. What is the best thing you can do?

GUIDELINES FOR OBSERVING COOPERATIVE GROUPS

1. The gender composition and the size of the group(s) they are observing, as well as what its task is.

2. Who is the group's leader, if there is one? How has this been decided?

3. How are the roles allocated among group members? How has this been decided?

4. Does everyone get a chance to do each role? If not, who (male or female) does which task?

5. Does everyone contribute equally to the group's work? If not, who (male or female) contributes most and who least?

6. Does everyone speak equally in the group? If not, who (male or female) speaks most and who least?

7. Does everyone have equal use of any equipment or other resources? If not, who (male or female) has most use and who least?

8. Does the teacher intervene in group functioning? Why, how, and with what results?

9. Do group dynamics differ according to whether they are single-sex or mixed-sex?

10. What techniques could be used by a teacher to prevent inequitable group dynamics and promote equitable ones?

TWO TECHNOLOGY VIGNETTES

A. You are teaching in a well equipped computer lab. Each student has his or her own computer and is working alone. You notice that many of the girls in the class seem to be bored. You know that girls are often happier working in groups, so you assign all your students to mixed-sex groups and ask them to continue their work collaboratively.

Is this approach likely to be successful? Why or why not?

B. You now notice that the girls in the groups you have assigned are the secretaries and errand-runners, while the boys are making the decisions and doing the actual work at the computers.

What do you do?

SCENARIO: MR. CHANG

Mr. Chang, a sixth grade teacher, has noticed that one of his female students, Sara, was beginning to give up working on mathematics problems if she did not get the answer on her first attempt. Most of the time, Sara's mathematics assignments are complete and she has the right answers, but Mr. Chang is noticing that Sara is becoming unwilling to try new approaches if she gets an incorrect response. This attitude is especially noticeable on open-ended tasks where strategy is important in getting the solution. He believes this is due to a lack of confidence in mathematics and Sara's belief that her ability is due to chance and not effort.

During the last lesson, Mr. Chang gave the class the following problem:

> Suppose five people meet at a family party and that each
> person shakes hands with each of the other people once.
> How many handshakes will there be?

Sara gave the answer of 15 handshakes.

When Mr. Chang asked Sara to explain how she arrived at the answer, Sara said that she made a lucky guess. Mr. Chang feels this is a critical time in Sara's mathematics education and wants to build Sara's confidence.

What could Mr. Chang do to help Sara realize that her ability to reason mathematically enables her to explore math activities and that the answers do not come by chance?

General Model of Achievement Choices

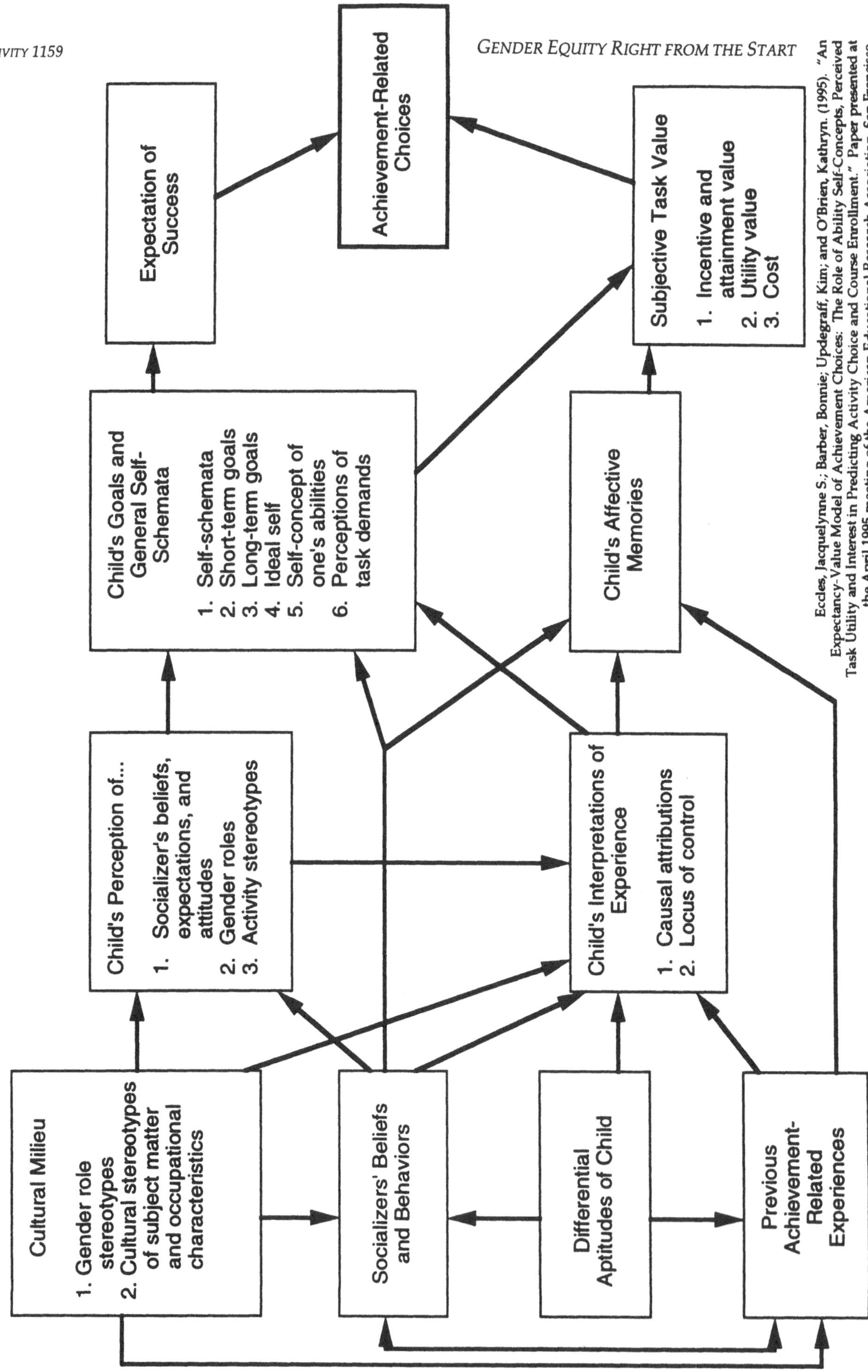

Eccles, Jacquelynne S.; Barber, Bonnie; Updegraff, Kim; and O'Brien, Kathryn. (1995). "An Expectancy-Value Model of Achievement Choices: The Role of Ability Self-Concepts, Perceived Task Utility and Interest in Predicting Activity Choice and Course Enrollment." Paper presented at the April 1995 meeting of the American Educational Research Association, San Francisco. Used by permission.

VIGNETTE: DAVID AND SUSAN

David and Susan are each working on focusing microscopes in tenth grade biology lab. Their lab stations are adjacent to each other. Susan is frustrated: She just cannot get the microscope to focus. David is having difficulty as well. They call the teacher, Mr. Bauer, over for assistance.

Mr. Bauer looks into Susan's microscope and focuses the image for her. "Thanks!" says Susan. Mr. Bauer then looks into David's microscope and says to him, "Just turn the fine adjustment slightly and you should get it focused." David works on it by himself.

What might Susan be thinking? What do you think David is thinking? Which student received the best help? Why? How is this related to learned helplessness?

AUTONOMOUS LEARNING BEHAVIOR MODEL

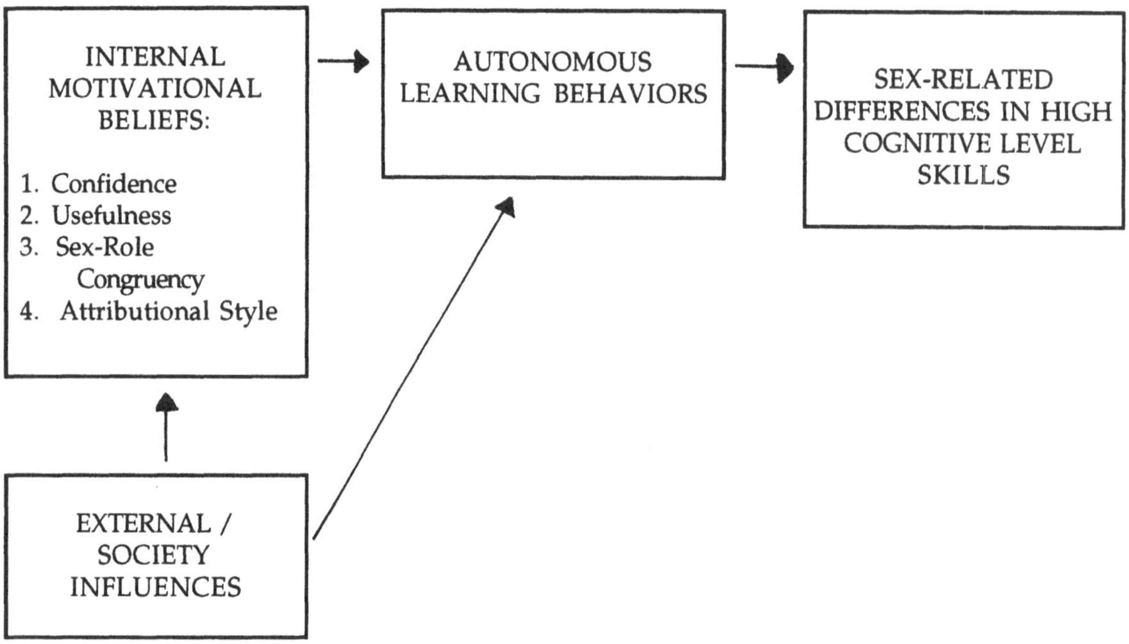

From Fennema, E. and Peterson, P. L. (1985). Autonomous learning behavior: A possible explanation of gender-related differences in mathematics. In L. C. Wilkinson and C. B. Marrett (eds.), *Gender-related differences in classroom interactions*, pp. 17-35. New York: Academic Press. Reprinted with permission.

VIGNETTE: THE A-V SQUAD

In your school, the A-V squad, which handles all the technology equipment including the TV, the VCR, and the laser video-disk player, is all male. You suggest to several girls in your class that they consider joining the squad.

"Oh, no," says Lakeesha, "that's just for boys." Lila agrees: "Girls don't know how to do those mechanical things. I sure don't!" "The boys would never let me touch that stuff!" chimes in Sarita.

What is the effect of these beliefs on girls? What would you do about this situation?

Looking through the Lens of Science

TIPS FOR TEACHERS by Janice Koch
Science Teams Teachers' Manual

All of us have been exposed to school science for varying numbers of years and with different levels of intensity. We probably all have many school science stories to tell! If we reflect on the entire span, from our earliest memories to the present, we can begin to construct our science autobiographies. The science autobiography is an attempt to touch a childhood school experience as seen through the eyes of the adult narrator. These autobiographies are a way of understanding our school science experiences and viewing them anew through our teacher lenses.

Try thinking back to your school science experiences from your earliest memory. What was science like for you? Did you have much of it in elementary school? Did you like it? Hate it? Rarely think about it? What happened in the middle grades? In high school? In college? What are your stories from school that helped to shape your attitudes and beliefs about science education?

As you reflect upon your own experiences, do you remember feeling like a science "insider" or a science "outsider?"

"If we are to achieve a richer culture, rich in contrasting values, we must recognize the whole gamut of human potentialities, and so weave a less arbitrary social fabric, one in which each diverse human gift will find a fitting place."
Margaret Mead
Sex and Temperament in Three Primitive Societies *(1935)*

Sheila Tobias, a researcher from the University of Arizona, used the "insider/outsider" terms to relate to mathematics learning, but they are also applicable to science learning. The science "insiders" are those who take for granted that they will do well in science and are willing to tackle school science *on its own terms*. Whatever the school science experience, insiders believe they can handle it. Typically, but not exclusively, science insiders are white males who have received media messages that present the image of scientists as white males.

Who are the science "outsiders?" These students frequently feel anxious during science class, are unsure about their ability to do science, and see the study of science as alien to themselves. Science outsiders have no sense of entitlement toward science and are not accustomed to seeing people like themselves portrayed as scientists by the media. When they do, those scientists are noted as exceptional for their gender, race, or ethnicity.

Here are some anecdotes recorded by college students who were asked to reflect on their science education. It is not hard to see the "insiders" and "outsiders" emerging from these anecdotes, and to recognize the harm done by stereotyped expectations.

Insider or Outsider?

"In seventh grade, my teacher asked me to come over to him to hold something and when I opened my hand, he deposited two cow eyes in my hand and laughed when I screamed, saying that all girls reacted this way."

"In tenth grade biology, I asked to sit out the lab on blood because I could not lance my finger. The teacher grabbed my hand and jabbed me in the finger with the lance."

"In elementary school, my teachers were annoyed by the many questions I would ask when we had science. They were much more interested in covering their lessons than in helping me to achieve personal knowledge."

"During junior high school science class, I dropped a test tube, but my teacher just told me to clean it up and he said it was okay for an experimenter to break something sometimes. That gave me more courage to ask questions and take risks in doing experiments."

"The sciences have been seen as masculine, not simply because the vast majority of scientists have historically been men, but also because the very characteristics of science are perceived as sex-linked."

Elizabeth Fee
International Journal of Women's Studies, *Volume 4*

"In first grade, I remember that we filled a pan with water and left it outside in the snow. When we checked it later in the day, we discovered that it was frozen. We read the temperature of the thermometer outside our classroom every day and checked our frozen water. It was wonderful!"

"My first experience with science took place in third grade. I had to do a science experiment. I did one on plants and it was a lot of fun. By sixth grade I did a science project on weather, entered a science fair, and won second prize. I grew up believing that (in science) we were active, learning, and discovering."

"In fourth grade, the science class was doing experiments with weight. One of the experiments involved using our own body weight. I was overweight at the time and this experiment made me feel very uncomfortable. The teacher had two of my peers weigh each person in the class. The outcome was embarrassing for me; I had fretted over it for days."

"My lab partner and I were having trouble locating the paramecium on the slide when our science teacher said, 'Come on girls, hurry up; if you can't find it ask the boys behind you.' I always believed that boys were 'cut out' for that sort of work."

"I was the only girl in my high school AP chemistry class and my teacher insisted on being my lab partner so that he could work with me. He always made remarks insinuating that if he let me do my own experiments, I would blow up the lab. He did my experiments for me."

"I have grown up to believe that most scientists are dull, boring people who spend most of their lives in a laboratory."

"When I entered the doctoral program in chemistry, I was struck by the absence of people of color—not only in any of the faculty—but in the entire student body! I was it—one of a handful of women and the only black person. I remember thinking, what am I doing here?"

What is Your Story?

How we construct our perceptions and interpret our experiences has important implications for our roles as educators. In science, it is particularly important to evaluate your students' needs and provide activities and a classroom environment that stimulates, encourages, and includes your students rather than shuts them out as "outsiders." In science education, it is important to evaluate where you have been, where you are now, and where you would like to go. Invite students to explore their assumptions about science as revealed by their own science stories.

"Like everyone else, scientists are born and raised in a particular culture of beliefs and biases and to one degree or another they will be affected in their work by what they believe, want, or need to be true."
Ruth Bleier (1984)

Try this activity yourself and then extend it to your students. First draw a picture of a scientist. Then consider the following: What is your scientist wearing? What physical characteristics does your scientist have? What is your scientist doing? What does *your* picture look like? Asking students to draw a scientist reveals that as early as second grade, students have deeply embedded beliefs about scientists. The dominant culture's stereotype of the scientist is adopted by the youngsters as their own. They imagine scientists as white men who have unusual hair, wear glasses and lab coats, and are involved with some sort of explosion. As one college student wrote:

"My eighth grade teacher proved my belief that scientists were weird men with glasses and hair that looked like they had stuck their finger in an electric socket."

This stereotype is limiting. It is not an accurate reflection of the existing scientific workforce and is certainly a distorted projection of the scientific workforce of the future.

Did you know that . . . ?

- In 1846, a young woman from Nantucket Island, Maria Mitchell (1809–1889), received a gold medal from the King of Denmark because she was the first person to discover a comet with a Danish telescope. She was asked some years later to head the Vassar College program in astronomy.

- In 1940, Navy Admiral Grace Murray Hopper invented COBOL, one of the first computer languages.

- The first person to examine the waterways of Massachusetts for possible industrial contaminants was the chemist Ellen Swallow Richards at the turn of the twentieth century.

- Ada Lovelace developed the first true calculating machine in the early 1800s.

"The task that seems of primary importance . . . is to convert science into a liberating and healthy activity; science with a soul which would respect and love its objects of study and stress harmony and communication with the rest of the universe. When science fulfills its potential and becomes a tool for human liberation we will not have to worry about women 'fitting in' because we will probably be at the forefront of this 'new' science."

Rita Arditti (1976)

- The first woman to be employed as a geologist for the U.S. Geological Survey was Florence Bascom. Called the Stone Lady, she developed the Bryn Mawr geology department at the beginning of the twentieth century. Dr. Bascom is known to have said of science that "the fascination of any search for the truth lies not in the attainment . . . but in the pursuit, where all the powers of mind and character are brought into play and are absorbed in the task . . . . One feels oneself in contact with something that is infinite and one finds joy in sounding the abyss of science and the secrets of the infinite mind."

- Rachel Carson's book *Silent Spring* raised the world's consciousness about the pollution of the environment in the 1950s.

- Rosalind Franklin's work with X-ray crystallography led to the elucidation of the structure of DNA in the 1950s. Most people believe she would have shared the Nobel Prize with Watson and Crick, but she died before the award was made.

What Do These Women Have in Common?

Marie Curie (1903) (1911)
Irene Joloit-Curie (1935)
Gerty Radnitz-Cori (1947)
Dorothy Crowfoot Hodgkin (1964)
Rosalyn Sussman Yalow (1977)
Barbara McClintock (1983)
Rita Levi-Montalcinio (1986)
Gertrude Elion (1988)

These are the "Laureates" in science in this century! They won the Nobel Prize in these fields:

Physics: Marie Curie
Chemistry: Irene Joliot-Curie, Marie Curie, Dorothy Crowfoot Hodgkin
Physiology and Medicine: Gerty Radnitz-Cori, Gertrude Elion
Medicine: Rosalyn Sussman Yalow, Barbara McClintock
Biology: Rita Levi-Montalcinio

Where Are the Silenced?

In 1875, the American astronomer, Maria Mitchell, addressed the Association for the Advancement of Women with the following thoughts:

> "In my younger days, when I was pained by the half-educated, loose, and inaccurate ways that women had, I used to say, 'How much women need exact science,' but since I have known some workers in science who were not always true to the teachings of nature, who have loved self more than science, I have now said, 'How much science needs women."

In fact, science needs more than just a few good women for a bright future in a world heavily dependent on scientific discoveries and innovations as a means to solve problems.

The Science Education Crisis

The need to educate all students to become literate in science and mathematics has taken on increasing importance in order for the United States to remain economically competitive in the global marketplace. National, state, and local efforts to assist students in science and mathematics achievement raises the following questions relating to the crisis in science education: Who is engaged and disengaged in the science classroom? Who continues to take science courses beyond those that are required? Who determines what and how the students will be taught in our science classrooms?

Having a History

Equity in science education has many meanings. Women and minority men have always done science. Learning who these scientists were and what they accomplished is empowering for young students. Historian Sara Evans of the University of Minnesota reminds us that, "Having a history is a prerequisite to claiming a right to shape the future." Teachers, parents, schools, and society must make sure that women and minority men are actively encouraged to take and persist in their science courses and to consider the choice of careers in science and engineering.

Listening for All the Voices

Strategies of encouragement and inclusion can be developed by teachers at all levels. In the classroom, teachers need to listen to the small, quiet voices of science outsiders in the face of frequently louder and more aggressive insider voices, especially in the upper grades. Both girls and boys must be actively engaged in conversations about science; both must be encouraged to envision themselves as future scientists with the ability to shape research agendas. Both groups need to know about scientists of both genders. At every school level, cultural norms inhibit girls and people of color from feeling a personal identification with science.

Don't Do It for the Girls!

Frequently science teachers do the laboratory experiment for their female students who ask for assistance while encouraging the boys to figure it out on their own. Far from helping these girls, the message communicated to them by these well meaning science teachers is that "you are not able to do it yourself." Science is seen as a male province where assumptions about female squeamishness and lack of mechanical aptitude preclude female participation.

Getting Messy Is Girl's Play

"We need to reconstruct our understanding of science in terms born out of the diverse spectrum of human experience."
Evelyn Fox Keller (1985)

The expectation of squeamishness of girls in the biology laboratory is an example of cultural stereotyping that discourages girls from fulfilling their potential in science. Some women have remarked that they were actually delighted to handle messy experiments in science, but, for appearances' sake, felt they had to act squeamish.

Despite the girlhood stereotypes that tell girls "not to get messy," it is the women who do the real "messing about" as they maintain the daily fabric of life. They are messing about with diapers, sicknesses, toilets, and the business of birth! What's for dinner? Although women are engaged in nightly dissections in the preparation of the evening meal, stories like this one reported by a teacher are all too common: "I did not mind dissecting the frog in seventh grade though that was not how I reacted...I screamed so that the boys would notice I was acting like a girl." The message for teachers is clear: Getting messy is one way of knowing the world, so help your students take the icks and yucks out of science!

The Masculine Image of Science

Practicing these strategies of encouragement is particularly challenging for science teachers since they must regularly confront societal attitudes that continue to interpret scientific activity as inherently masculine. Girls are often given the impression that they can be feminine or scientific, but not both. The media contribute to this message by reinforcing stereotypes. When describing Nobel Laureate Rosalyn Yalow, the headline read, "She cooks, she cleans, she wins the Nobel Prize!" The article describing 1983 Nobel Laureate Barbara

McClintock begins by extolling her love of baking, as if to prove that she is a "real" woman despite her scientific talent.

Making Connections With the Real World

Young students value science more when their identification to the real world and people are emphasized. Too often science seems an arcane activity reserved for ivory tower eggheads. Actually, science is one form of creative human endeavor that can be a viable option for many people, not only the highly gifted. Science needs to be presented as a way of knowing about the world; it is essentially a social and often a cooperative activity with rich interpersonal as well as intellectual rewards. Visiting scientist programs help to illustrate that scientists are, indeed, real people. Local chapters of the Association for Women in Science can send women from all fields of the natural sciences to schools to talk to youngsters about "doing science." Many companies and universities are happy to provide role models for females and students of color.

Who Will Do Science?

Shirley M. Malcom, of the American Association for the Advancement of Science, remarks that there is a vast untapped talent pool comprising girls and young women and men of all races including Blacks, Hispanics, and American Indians, who are capable of doing science at every level. So far, science education has largely failed these groups. These are the students who receive the message from the dominant culture that science is not for them. These are the students we need to encourage through equity strategies.

An Agenda for Action

Teachers need to recognize why girls and minorities avoid science in order to begin understanding their reluctance to pursue science as a career. The reasons are found in the ways in which students are socialized, and the expectations that teachers and the community have for their achievement and their career plans. Since these students are alienated from science because of social pressures, educators need to compensate for the absence of role models and the lack of encouragement by media and society at large by making their science classrooms more receptive and inclusive.

Here are some guidelines to help you rate your teaching practices in science:

1. Do you call on boys more frequently than girls? Do you recognize or call on boys when they call out or appear likely to disrupt while expecting girls to wait patiently to be recognized? Do you include the quiet girls who do not volunteer?

2. Who are your four top students in science? What is their sex? Ethnicity?

3. Is your classroom sex or race segregated in seating, work groups, or informal interactions?

4. Do you use gender-free, bias-free language and occupational titles?

5. Do you make connections between scientific concepts and the outside world?

"Science is a way of thinking much more than it is a body of knowledge. Its goal is to find out how the world works, to seek regularities there may be, to penetrate to the connections of things . . . ."

Carl Sagan

6. Do you invite women and men of color in science to speak to your students and do experiments with them?

7. Do you do the science lab activity for the girls while encouraging the boys to do it on their own?

8. Do you have the same expectations for the achievements of your girls and minority boys in science as you do for the white boys?

9. Do you let all the students know that they are capable and competent in science?

10. Do you praise the girls for their appearance or the appearance of their work while praising the boys for their accomplishments?

Are YOU a Science Outsider?

The above questions are designed to begin a process of soul searching that may enable you to change those classroom behaviors that may inadvertently turn some students off to science. Teachers frequently teach as they have been taught, and it is necessary to break the chain of this disempowerment for many of our science outsiders. Many teachers realize that they have been science outsiders and work very hard on their own science insider journey through inservice course work like SCIENCE TEAMS. Myra and David Sadker (see Griffin, 1990), professors at American University, remind us that "girls are the only group who enter school scoring ahead and twelve years later leave school scoring behind. The decline of academic achievement of half our population remains an invisible issue." Shirley M. Malcom informs us that "the next generation of science must necessarily draw on young people who are not generally seen (and often do not see themselves) in the present mix. Who will do science? That depends on who is included in the talent pool."

References

Malcom, S. (1990). "Who Will Do Science?" *Scientific American*, February, 1990.

Griffin, K.R. (1990). "Attending to Equity." *AAUW Outlook*. American Association of University Women: February/March, 1991.

Rossiter, M. (1982). *Women Scientists in America: Struggles and Strategies to 1940*. Baltimore: Johns Hopkins University Press.

The AAUW Report. (1992). *How Schools Shortchange Girls*. AAUW Educational Foundation and National Education Association.

Tobias, S. (1988). "Insiders and Outsiders." *Academic Connections*, The College Board Publications, Winter, 1988.

This chapter is reprinted by permission from the Science Teams Teacher's Manual by Aleta You Mastny et al. For information about the manual, training, or video, contact Dr. Mastny at the Consortium for Educational Equity (see Appendix B).

ASSESSMENT PROFILE

Name of assessment
State/city
Year began
Grades administered
Type of assessment
How the assessment addresses the relevant education standard

WRITING PROMPTS

Janice Koch and Carolyn Sears

Presented at the American Educational Research Association
Chicago, April 1991

WRITING PROMPTS FOR A NEW KIND OF LAB REPORT

Pose challenges in science education. Invite everyone's participation, collect everyone's data and come to conclusions based on collective resources. Typical school laboratory activities are designed to yield the same responses from all students. They are contrived exercises with predictable results. "Did I get the right answers?" and "Can I copy your lab report?" are familiar questions in school science lab. A revision of laboratory reporting would legitimize the personal involvement in doing science by asking for individual comments about process and results. The format might resemble the following.

"My name is ________ . I worked with _________ on this lab in order to find out ______ . This is what we did: _________. This is what actually happened: _____________. The parts of this process that we were unhappy with were: ___________. As a result of our conclusions, we reasoned that ___________."

Introducing the first person to the content of the lab report connects the knower to the known, identifies the knower's agenda, and reminds students of the human nature of doing science.

SPECIAL-PURPOSE WRITING PROMPTS

1. Experimental Design. What will we measure? How will we do the lab?

Prompted by an anomaly or discrepant event, students begin by identifying the compelling question or the probable hypothesis. Next students identify the components of the event. *What is happening here?* Then they can identify what can be measured, design a data chart, and plan the lab. Different parts of the process can be used as writing prompts. Teaching the scientific method needs to be done over a long period of time and not in the order presented in formal lab report writing. As the teacher, I focus on one part of the scientific method at a time, and carefully build to an understanding of each part of the process.

2. I see considering an alternative explanation (hypothesis), learning to contradict yourself from day to day, and staying open to learning new ways of explaining an event as critical skills for students.

I used to think but now I think ...
I used to think but now I know ...
Be inventive: how else can the falling apple be explained?
When you feel that you don't know, try out two ideas: it can, and it cannot

3. To use data to substantiate ideas through collaboration or to expand ideas, interview three other students at the end of a lab and quote them in the journal. How does careful lab technique help you to develop a voice of confidence?

4. Prompt for process writing (metacognitive):

   To answer the question, here's how my thinking is going / went ________
   To make the prediction, my thinking was ____________

5. For collaborative work habits:

   Describe the tasks you did for the group.
   What could you have done to make your group work better?
   What worked well in your group?
   My plan for what I will do tomorrow is ___________

STUDENT ASSESSMENT OF RESEACH NOTEBOOKS

In portfolio assessment procedures, students are asked to describe the purpose of the document and the evidence of its effectiveness. The research notebook is a significant document for authentic assessment of student work in the sciences.

Daily writing prompts:

Today what have you learned about research? Experimental design? Science?
Today I learned ...
What have you learned about yourself and others while doing this lab?
If you encountered any problems, how would you avoid them next time?
If you had no problems, what advice can you give?

Periodic assessment (at five-week intervals):

What do you notice about your work? What did you write in your journal? How have you changed as a scientist? What are your goals for next quarter?
What is your best piece of research? What makes it best? What does one have to do in order to become a successful scientist?
What is the most important thing you learned about science this quarter? Why is your journal a demonstration of learning? How does the journal illustrate what you have learned?
What is the purpose of your journal and what is the evidence that you have achieved this purpose? How would you revise your statement of the purpose of your journal?

Campbell, Patricia B. (1989). *The Hidden Discriminator: Sex and Race Bias in Educational Research.* Newton MA: WEEA Publishing Center

A Beginner's Guide to Educational Research

What Is Research?

Research, according to Webster's, is an "investigation or experimentation aimed at the discovery and interpretation of facts." In education there are two major types of research: *basic research*, in which the goal is a better understanding of learning and the educational process, and *applied research*, which focuses on finding information that will improve current educational practice.

The Research Method

Educational research, like research in other areas, relies heavily on what is known as the scientific method. This is a system of investigation that typically involves the following steps:

- development and statement of a question or a problem to study
- formulation of a hypothesis (a "best-guess" answer to the question being studied, based on existing theory and research)
- development and implementation of a structured plan (or design) to test the accuracy of the hypothesis or to answer the questions posed
- determination of the results of the plan
- generation of conclusions based on research results and the development of further research questions based on both the results and the conclusions

The Sample

In research, those being studied are called participants, or *subjects*. A group of subjects is called the *sample*. The sample is supposed to be representative of a *population*, a larger group to whom the results of the research can be applied (i.e., the results of a study on a sample are generalized to the population that sample represents). For example, if you selected ten students from each of ten classes for a research project, each of the students would be a subject, the one hundred students would be your sample, and the ten classes would be the population to which your results could be applied.

The Design

Both basic research and applied research can be done using one of several different designs, or plans of action. These plans are based on a variety of factors, including the type of research being done and the resources available.

The Experimental Design. If one wishes to determine whether something "causes" something else, the experimental design is the most effective. In this design, one or more groups, called the *experimental group(s)*, receive some sort of treatment (e. g., a new reading program, a new tutoring program, or smaller class size) while a similar group, the *control group*, receives no treatment. The essence of the experimental design—that which makes it the "best" design for causal research—is that each of the subjects studied has an equal chance of being selected for either the experimental or the control group. Use of this design increases the chance that the only difference between the experimental group and the control group will be that one receives the treatment and one doesn't. Thus if differences show up between the experimental and control groups, those differences can be said to be caused by the treatment. For example, if the one hundred students in our sample are selected by chance to go either into a group that receives money for getting an A or into a group that receives no money for getting an A, then we have an experimental design. If the subjects receiving money get higher grades than those who don't receive money, then, because it is an experimental study, we can say that, for that sample, receiving money for good grades improves student grades.

The Ex Post Facto Design. Experimental designs are not always appropriate. For instance, because people cannot be randomly assigned to be female or male or to be Black or white, an experimental design cannot be used to study sex or race. Neither can an experimental design be used to study something that has already occurred, because it is then too late to randomly assign subjects.

For these kinds of studies an *ex post facto*, or *quasi-experimental* design can be used. *Ex post facto* is a Latin expression meaning "after the fact." In an ex post facto study, the researcher does not control who is in the experimental group and who is in the control group. Therefore, it is not possible to be sure that the only difference between the groups being studied is the treatment or to conclude that the treatment "caused" any differences in the group.

For example, a researcher studying young children at play might find that girls and boys have different patterns of play. The researcher could conclude that girls and boys play differently but could not conclude that being a girl or being a boy "caused" the children to play differently. There are many other variables that might account for the differences. These variables might include that all the boys are wearing pants while a number of the girls are wearing dresses; that teachers frequently give different instructions and play suggestions to girls versus boys; or that parents are more apt to be concerned about girls "keeping clean" than about boys doing so.

Post Hoc Fallacy. Drawing invalid conclusions based on ex post facto research is so prevalent that researchers have a special name for it: a *post hoc fallacy.* Since by definition all research comparing females with males, people of color with whites, and disabled with able-bodied persons is ex post facto, it is particularly important to check for post hoc fallacies in such studies. You should suspect the existence of a post hoc fallacy whenever a study concludes that being female or Black or disabled, for example, causes something to happen, whatever that something might be.

Other Designs. There are a number of other ways that research can be done, including the following:

- *Survey, or descriptive, research,* in which there is no treatment and subjects respond to a series of written or oral questions describing a situation or area of interest. A study of student attitudes toward school would be an example of survey research.
- *Qualitative, or naturalistic, research,* in which the researcher observes people in a natural setting and, over a period of time, almost becomes a part of a group in order to be able to analyze group processes and interactions. A study of how fourth-graders' behavior changes in terms of how fourth-graders interact with the teacher during the school year would be an example of a qualitative study.
- *Correlational research,* in which the degree to which changes in one variable are reflected in changes in one or more other variables. A study of the relationship between achievement test scores and grades would be an example of correlational research.
- *Historical research,* in which analysis is based on documents and data from the past. A study of the different ways that reading was taught in the nineteenth century would be an example of historical research.

It is important to note that regardless of the research design used, *if there is no random assignment of subjects to the treatment, then you cannot be sure that the treatment caused any differences.*

Sources of Invalidity

Obviously, the quality of research—its validity, the degree to which results are accurate and can be attributed to that which is being studied—is very important. Researchers have long been concerned about the validity of their work and have

attempted to design studies that control for as many sources of invalidity as possible, even though few control for or even consider societal biases. The following is a list of the more common sources of invalidity not related to societal biases.

- *The Hawthorne Effect.* Being studied and getting extra attention, being "special," may be enough to cause changes in the subjects independent of what is being studied. The original Hawthorne study was done with factory workers. Researchers found that productivity increased when they did positive things (increased light, increased breaks); they also found that productivity increased when they did negative things (increased room temperature). Further work found that it was the increased attention that raised productivity. Using a second group that gets the attention but not the treatment controls for the Hawthorne Effect.
- *Maturation.* Just growing older can have a strong influence on subjects, particularly if young children are being studied. For example, if researchers are studying the effects of a year-long program on children's language development, they must remember that children's language skills will improve in a year regardless of the program used. Without a same-age control group, the researcher will not know how much of a change is caused by that which is being studied and how much is caused by the subjects' getting older.
- *Testing.* Testing can affect a study in many ways. Obviously, if a test doesn't measure what it is supposed to measure, then results will be incorrect. In addition, taking a test can affect subjects; changes in subjects may be due to the test rather than the treatment. For example, the practice of taking a pretest on fractions might do more to increase students' abilities to work with fractions than the treatment does. Finding tests that have been found to be valid (that do measure what they say they measure) and using a control group that takes the tests but not the treatment are ways of controlling for the influence of testing.
- *History.* In an ideal study, the only difference between groups being studied is the treatment; however, during a study, groups may have different experiences (a teacher might get sick, a school might start a new project). By the end of the research period, the different histories of the groups, rather than that which is being studied, might be the cause of any changes. It is very difficult to control for history; being aware of the unexpected and unintended events that occurred and reporting them in the results are about all that can be done.

Statistics

Wait! Even though your first impulse may be to skip this section, *read on.* Many of us are afraid of statistics and convinced that we can never understand them. That does not have to be the case. Even though most of us will never become theoretical or even applied statisticians, we can, with a little effort, learn enough to begin to make sense of the statistical section of a piece of research. Always keep in mind that statistics are just a way of reducing large amounts of information (test scores, height, attitudes, rankings, almost anything) into summaries that provide useful information.

There are two basic types of statistics: *descriptive* and *inferential*.

Descriptive Statistics

Descriptive statistics as their name implies, reduce and describe a large amount of information. Typical descriptive statistics include the *mean* (average), *mode* (most frequent score), and *median* (score at which half the scores are below and half are above). The *standard deviation* is the measure of how varied or spread out a set of scores is. If a set of scores has a mean of 10 and a standard deviation of 1, most of the scores in that set will be very close to one another and to the mean of 10. About two-thirds of the scores will be between 9 and 11. A group whose scores are close together is called *homogeneous*. A group in which the scores are much more spread out, where the standard deviation is much larger, is called *heterogeneous*. For example, a set of scores with a mean of 10 and a standard deviation of 5 is more spread out than the first example (it is therefore heterogeneous); about two-thirds of this group's scores would be between 5 and 15.

Other descriptive statistics include *stanines*, *percentiles*, and *standard* scores. Stanines break the distribution of scores into nine sections (1=lowest, 9=highest) and indicate in which of the nine sections an individual score falls. Percentiles (from 0 to 99) describe the percentage of scores that are lower than an individual score. Standard scores describe how far an individual score is from the mean score: if 0 represents the mean, then a standard score of 1.5 means that the individual score is one and one-half standard deviations above the mean, whereas a score of minus 2 means a score is two standard deviations below the mean.

Inferential Statistics

Whereas descriptive statistics describe what the information is, *inferential statistics* tell what can be implied from that information. Inferential statistics tell us the odds in which the differences between groups can be attributed to chance or are real and could be replicated (that is, if the study were done again, the results would be similar).

Most researchers feel that they have to be at least 95 percent certain that differences between groups are real before they are willing to say so. This is known as the *level of probability* or *significance* and is generally shown as $p < .05$, meaning that the differences between groups were large enough that the chances are better

than 95 out of 100 that the study could be replicated with similar results. The statement ($p < .05$) is considered an acceptable level of risk, and differences between groups at the $p < .05$ level are considered statistically significant. If the study reports ($p < .01$), then the chances are better than 99 out of 100; ($p < .001$) means the chances are better than 999 out of 1000; and so on.

Significant Differences

There are two types of significant differences: *statistical* and *practical*. Statistical significance means that the differences between or among groups are most likely real and would be found if the study were replicated. However, just because a difference is statistically significant does not necessarily mean that it has practical meaning. Statistical significance is related to a number of things, including the size of the differences between groups, the number of subjects, and the degree that the scores in each group are spread out. For example, a study of 10,000 students might find that students using Math Book A increased their math achievement 1 percent more than students using Math Book B did. This difference would be statistically significant, meaning that the differences were most likely real and not due to chance. A teacher choosing a math book, however, could and most likely would say that a 1 percent difference was not meaningful; with such a small difference, factors such as cost and ease of use would and should be the deciding factors in book selection. Thus the statistically significant difference would have no practical significance.

This has been only an introduction to research methods; there is much more to learn. For additional information, see F. Kerlinger's *Foundations of Behavioral Research* (New York: Macmillan, 1972) or any of the many educational research books available in local college or public libraries.

EVALUATION AND DOCUMENTATION, OR, WHY SHOULD I DO THIS?

Patricia B. Campbell, PhD
Campbell-Kibler Associates, Inc.

WHY EVALUATE?

1. To determine the effects the program is having.
2. To identify program strengths and weaknesses, and develop strategies to improve weak areas.
3. To show funders, potential funders, and other supporters what the program does and how it is working.

HOW EVALUATION CAN HELP

1. Provide information about program activities for use in reports and funding proposals.
2. Provide periodic feedback on how well your program is going.
3. Other ways that are unique to your program.

LEVELS OF EVALUATION

The **monitoring level** focuses on documenting what was done, answering such questions as "What are you doing?" and "Are you doing what you said you would?"

The **quality level** focuses on information that can be used to improve the on-going program, answering such questions as "Is the work being done well?" To some extent all the information collected is used at this level with a particular emphasis on observations and interviews.

The **effectiveness level** focuses on the effects of the program on individuals and institutions. It answers such questions as "What effect, if any, did the program have on young women?"

EVALUATION FIRST STEPS

1. **Decide the specific questions you want to answer.**
   One question always asked is: "Is the program a success?" However, before you can begin to answer it you need to decide what it would take before you (and others) would see your program as a success.

2. **Decide if you want the evaluation to be done by you or an outside professional evaluator.**
   It is frequently better to do the evaluation jointly so you can call on the expertise and objectivity of the evaluator while keeping cost down and relying heavily on inside program knowledge.

3. **Decide on the type of evaluation you want to have.**
   For example, you could focus on documenting what was done, the collection of information that can be used to improve the on-going program, the effects of the program on individuals and institutions, or some combination of these.

4. **Decide on the information you want to collect.**

5. **Get started.**

CAN EVALUATION BE USEFUL AND EVEN FUN?

Patricia B. Campbell, PhD
Campbell-Kibler Associates, Inc.

To most the answer to the above question is a resounding No! When one is using traditional methods, evaluation can sometimes be useful, but is rarely fun. The methods to answer the question "How's it going?" tend to be guesswork, observation, and participant ratings on a scale of 1 to 5, on a variety of factors. These methods don't provide you with useful, timely information and aren't even fun. However, some methods can be both useful and fun.

ON THE BEAN

Take three different kinds of beans (e.g., pinto, kidney, and black), or colors of buttons or poker chips. Label their containers one GREAT (or a smiling face), the second OKAY (or a neutral face), and the third AWFUL (or a sad face). Have participants choose a bean based on how they feel about what's happening and put it into a clear glass jar. By looking at the jar you can tell immediately "how it's going."

WORDS, WORDS, WORDS

Ask participants to list three words that best describe how they feel. By examining the most frequently listed words, you can get an idea of how participants feel about "what's happening" and its impact.

THE BIG THREE

Ask participants to list what they LIKED BEST and LIKED LEAST about the session and what would IMPROVE IT. Reading or summarizing the most frequent responses will give an overview of their feelings.

I THINK ...

Ask participants to complete one or several of the following. I THINK TODAY'S SESSION IS ..., TODAY I LEARNED THAT ..., IN THE FUTURE I WILL .... These can be done on paper or, if you prefer, on easel paper taped to the wall. In the latter case, attach a marker to the wall with a string next to the paper. At the top of the sheet write the sentence prompt.

LET'S TALK ABOUT IT

The last three activities ("Words, Words, Words," "The Big Three," and "I Think ...") can be done as small group activities instead. In groups of three or four, with one person serving as the recorder, participants discuss their feelings and ideas. The recorders can report back to the large group or they can hand in their notes.

SOME FOR ME, SOME FOR YOU

If there are a number of evaluation questions to be answered, randomly give half the questons to half the participants and the other half to the rest.

These are just a beginning. Try them and see how they work for you, or use them to develop your own ideas.

SAMPLE GENDER **ATTITUDE** SURVEY QUESTIONS

Patricia B. Campbell, Ph.D.
Campbell-Kibler Associates, Inc.

Important Note

Although attitudes are important, knowledge and behavior — real impact on students — are what counts. Teachers (and professors) may have positive attitudes, but there will be no impact on students if the teacher does not know what to do or if the teacher knows what to do but doesn't do it.

Today in many places, positive attitudes toward gender equity are "socially acceptable." Thus it is difficult to determine if responses reflect people's attitudes or reflect what is seen as the "right" answer. This is particularly serious when students are filling out attitude questionnaires for those who grade them. For these reasons as well as issues of validity and reliability, attitude surveys should not be the only measure used in any evaluation. Along with measures of behavior and knowledge, however, they can give some valuable information. The following are some sample attitude questions you can use in a survey.

1. How important is it to make a special effort to encurage girls to participate in math/science/technology?

1	2	3	4	5
Very important				Not at all important

Do you do anything special to encourage girls?

___ Yes ___ No

Please describe what you do.

2. Why do you think there are more men than women in science and engineering careers?

3. Please complete the following sentences any way you want.

When I do math/science/technology, I feel ...

Girls who do math/science/technology are ...

A girl has the right to ...

Boys who do math/science/technology are ...

Boys have the right to ...

4. To complete the following, please place an X in the space between each pair of adjectives that best describes your feelings about GENDER EQUITY IN EDUCATION.

GENDER EQUITY IN EDUCATION

GOOD	___:___:___:___:___:___:___	BAD
NICE	___:___:___:___:___:___:___	AWFUL
CRUEL	___:___:___:___:___:___:___	KIND
BORING	___:___:___:___:___:___:___	EXCITING
HAPPY	___:___:___:___:___:___:___	SAD
CLEAN	___:___:___:___:___:___:___	DIRTY
SAFE	___:___:___:___:___:___:___	DANGEROUS
HARD	___:___:___:___:___:___:___	EASY
IMPORTANT	___:___:___:___:___:___:___	UNIMPORTANT

5. What do you feel are the major goals of gender equity efforts in math/science/technology?

6. What do you feel are the major barriers to achieving gender equity in math/science/technology?

SAMPLE GENDER **BEHAVIOR** SURVEY QUESTIONS

Patricia B. Campbell, PhD
Campbell-Kibler Associates, Inc.

1. What is the average amount of time you devote to covering gender equity in each of your methods courses each semester/quarter?

   ___ About 2 hours ___ Between 1 and 2 hours ___ Less than 1 hour ___ None

   ___ Other: please specify

   ----

2. Please check each of the gender equity topics you cover in your methods course.

   ___ The underrepresentation of women in math/science/technology careers

   ___ The underrepresentation of girls in advanced math/science/technology courses

   ___ Stereotypes associated with math/science/technology

   ___ Teacher/student interaction patterns favoring boys

   ___ Identification of gender bias in the curriculum

   ___ Title IX and other laws relating to gender bias and gender equity

   ___ Specific how-to strategies for increasing girls' parrticipation in math/science/technology. Which strategies?

   ___ Other: Please specify

   ----

3. If you have incorporated gender equity materials and activities in your classroom, what impact, if any, has it had on:

   A. What you teach?

   B. How you teach?

   C. The students?

D. You as a professional?

4. Please check any of the following gender equity activities you do.

___ Read literature about gender equity

___ Publish books or articles about gender equity

___ Attend conferences or sessions about gender equity

___ Present sessions on gender equity

___ Discuss gender equity with colleagues

___ Other: Please specify

WORKSHOP PROGRAM EVALUATION

Name ____

School/College/University _________________________ Date __________

Subjects/Courses taught ____

1. What aspects of this program will be MOST helpful to you in the future?

2. What aspects of this program will be LEAST helpful to you in the future?

3. In what ways, if any, are you planning to change your teaching based on this program?

4. Please list up to three things you learned from this program that you feel will affect your professional work, including your teaching.

1.____

2.____

3.____

Patricia B. Campbell, PhD
Campbell-Kibler Associates, Inc.

WORKSHOP FOLLOW-UP

Name ____

School/College/University____

1. In what ways, if any, do you incorporate gender equity in your classes?

2. Are you planning to make any changes in the way you cover gender equity?

   ___ Yes ___ No

   What changes are you planning to make?

3. How do students react to the gender equity areas of your courses?

4. In what ways could you improve your coverage of gender equity in your classes?

5. In what ways, if any, did attending the gender equity session influence:

   The gender equity topics you cover?

   How you cover gender equity?

Patricia B. Campbell, PhD
Campbell-Kibler Associates, Inc.

RESOURCES

READINGS FOR ACTIVITY 1, REVIEWS OF THE RESEARCH

M = MATHEMATICS S = SCIENCE T = TECHNOLOGY

S Antony, Mary (1993-94). Gender and science: A review of the research literature. *Equity Coalition Newsletter,* 3 (2) , 10-13.

S Benditt, John (Ed.). (1992). Women in science: First annual survey. *Science,* 255, 1365-1388.

M Campbell, Patricia B. (1991). Girls and math: Enough is known for action. Newton, MA: *Women's Educational Equity Act Publishing Center Digest,* June issue, Education Development Center.

M, S CEW Research Reports (1992). Studies of determinants of women's achievement in mathematics and science. *Women in mathematics and physics: Inhibitors and enhancers.* Ann Arbor, MI: Center for the Education of Women, University of Michigan.

S Culotta, Elizabeth & Gibbons, Ann (Eds.). (1992). Minorities in science: The pipeline problem. *Science,* 258, 1175-1235.

M Davenport, Linda Ruiz (1994). Promoting interest in mathematical careers among girls and women. Columbus, OH: *The Mathematics Outlook,* October ERIC/CSMEE bulletin. Columbus: OH.

M Fennema, Elizabeth & Leder, Gilah C. (1990). *Mathematics and gender.* New York: Teachers College Press.

T Fisher, Allan, Margolis, Jane, & Miller, Faye (1997). Undergraduate women in computer science: Experience, motivation and culture. Proceedings of the Association of Computing Machinery's Special Interest Group on Computer Science Education Technical Symposium, February 1997.

M Hanson, Katherine (1992). Teaching mathematics effectively and equitably to females. Newton, MA: Center for Equity and Cultural Diversity, Working Paper #1.

S Harding, Sandra (1986). *The science question in feminism.* Ithaca, NY: Cornell University Press.

M Hyde, Janet S. et al. (1990). Gender comparisons of mathematics attitudes and affect. *Psychology of Women Quarterly,* 14, 299-324.

S, M Kahle, Jane Butler & Meece, Judith (1994). Research on girls in science: Lessons and applications. In Dorothy Gabel (Ed.), *Handbook of research in science teaching and learning* (pp. 542-557). Washington, DC: National Science Teachers Association.

M, S Linn, Marcia C. (1992). Gender differences in educational achievement. *Sex equity in educational opportunity, achievement, and testing.* Proceedings of the 1991 invitational conference of the Educational Testing Service, Princeton NJ.

M, S Linn, Marcia C. & Hyde, Janet S. (1989). Gender, mathematics and science. *Educational Researcher*, November, 17-27.

M, S Linn, Marcia C. & Petersen, Anne C. (1986). A meta-analysis of gender differences in spatial ability: Implications for mathematics and science achievement. In J. S. Hyde & M. C. Linn (Eds.), *The psychology of gender: Advances through meta-analysis* (pp. 67-101). Baltimore: Johns Hopkins University Press.

T Mark, June (1992). Beyond equal access: Gender equity in learning with computers. *Women's Educational Equity Act Publishing Center Digest*. Newton, MA: Education Development Center.

T Neumann, Delia (1991). Technology and equity. *ERIC Digest*, (ED339400). December 1991.

M, S Parker, Lesley H., Rennie, Léonie & Fraser, Barry J. (Eds.). (1996). *Gender, science and mathematics: Shortening the shadow.* Dordrecht, The Netherlands: Kluwer Academic Publishers.

M Parker, Marla (Ed.). (1995). *She does math: Real-life problems from women on the job.* Washington, DC: The Mathematical Association of America.

S Rayman, Paula & Brett, Belle (1993). *Pathways for women in the sciences.* Wellesley, MA: Center for Research on Women.

M Rogers, Pat & Kaiser, Gabriele (Eds.). (1995). *Equity in mathematics education: Influences of feminism and culture.* Washington, DC: Falmer Press.

T Sanders, Jo (1986). *The neuter computer: Computers for girls and boys.* New York: Neal-Schuman Publishers.

M Schwartz, Wendy & Hanson, Katherine (1992). Equal mathematics education for female students. *ERIC/CUE Digest*, 78, (ED344977).

M Secada, Walter G., Fennema, Elizabeth, & Adajian, Lisa B. (Eds.). (1995). *New directions for equity in mathematics education.* New York: Cambridge University Press.

M Secada, Walter G. & Meyer, Margaret R. (Eds.). (1989). Needed: An agenda for equity in mathematics education. *Peabody Journal of Education*, 66(2).

T Sutton, Rosemary E. (1991). Equity and computers in the schools: A decade of research. *Review of Educational Research*, 61(4), 475-503.

S Weinburgh, Molly (1995). Gender differences in student attitudes toward science: A meta-analysis of the literature from 1970 to 1991. *Journal of Research in Science Teaching*, 32 (4), 387-398.

S, M Wilson, Meg (Ed.) (1992). *Options for girls: A door to the future: An anthology on science and math education.* Austin, TX: Pro-Ed.

M Wisconsin Center for Educational Research (1994). Equity and mathematics reform. *NCRMSE Review: The Teaching and Learning of Mathematics.* Madison, WI: University of Wisconsin.

RESOURCES FOR OTHER ACTIVITIES

American Association of University Women Foundation (1992). *How schools shortchange girls.* Washington, DC: American Association of University Women.

American Association of University Women Foundation (1995). *How schools shortchange girls.* New York: Marlowe & Co.

Angier, Natalie (1991). Women swell ranks of science, but remain invisible at top. *New York Times,* May 21, 1991.

Bornholt, Laurel J., Goodnow, Jacqueline J., & Cooney, George H. (1994). Influences of gender stereotypes on adolescents' perceptions of their own achievement. *American Educational Research Journal,* 3 (31), 675-692.

Brunner, Cornelia (1991). Gender and distance learning. In L. Roberts & V. Horner (Eds.), *Annals of the American Academy of Political and Social Science.* Beverly Hills, CA: Sage Press.

Burns, Marilyn (1977). *The good times math event book.* Worth, IL: Creative Publications.

—— (1987). *A Collection of math lessons.* San Francisco: The Math Solution Publications.

Caduto, Michael & Bruchac, Joseph (1994). *Keepers of life: Discovering plants through Native Americans stories and earth activities for children.* Golden, CO: Fulcrum Publishing.

Clement, Frank Rod (1991). *Counting on Frank.* Milwaukee ,WI: Gareth Stevens.

Clewell, Beatriz Chu, Anderson, Bernice Taylor, & Thorpe, Margaret E. (1992). *Breaking the barriers: Helping female and minority students succeed in mathematics and science.* San Francisco: Jossey-Bass Publishers.

Davis, Cinda-Sue & Rosser, Sue (1996). Program and curricular interventions. In Cinda-Sue Davis, Angela B. Ginorio, Carol S. Hollenshead, Barbara B. Lazarus, Paula Rayman & Associates (Eds.). *The Equity Equation: Fostering the advancement of women in the sciences, mathematics and engineering* (pp. 232-264). San Francisco: Jossey-Bass Publishers.

Dillow, Karen, Flack, Marilyn, & Peterman, Francine (1994, November). Cooperative learning and the achievement of female students. *Middle School Journal,* 48-51.

Dumond, Val (1990). *The Elements of nonsexist usage: A guide to inclusive spoken and written English.* New York: Prentice Hall Press.

Eccles, Jacquelynne (1987). Gender roles and women's achievement-related decisions. *Psychology of Women Quarterly,* 11, 135-172.

—— (1989). Bringing young women to math and science. In Mary Crawford & Margaret Gentry (Eds.), *Gender and thought: Psychological perspectives.* New York: Springer Verlag.

—— (1994). Understanding women's educational and occupational choices: Applying the Eccles et al. model of achievement-related choices. *Psychology of Women Quarterly*, 18 (4), 585-625.

——; Barber, Bonnie; Updegraff, Kim; & O'Brien, Katherine M. (1995). *An expectancy-value model of achievement choices: The role of ability self-concepts, perceived task utility and interest in predicting activity choice and course enrollment.* Paper presented at the annual meeting of the American Educational Research Association, April 18, 1995, San Francisco.

Eccles-Parsons, Jacquelynne, Meece, Judith L., Adler, Terry F., & Kaczala, Caroline M. (1982). Sex differences in attributions and learned helplessness. *Sex Roles*, 8, 421-432.

Fryer, Bronwyn (1994). Sex and the superhighway. *Working Woman*, 19 (4), 51-62.

Gega, Peter (1993). *Concepts and experiences in elementary science.* New York: Macmillan.

Greenberg, Selma (1978). *Right from the start: A guide to nonsexist child rearing.* Boston: Houghton Mifflin Co.

Greenberg, Selma (1985). Equity in early childhood environments. In Susan S. Klein (Ed.). *Handbook for achieving sex equity through education* (pp. 457-469). Baltimore: Johns Hopkins University Press.

Grossman, Herbert and Grossman, Suzanne (1994). *Reducing gender-stereotypical behavior: Gender issues in education.* Needham Heights, MA: Allyn and Bacon.

Guild, Pat (1994). The culture/learning style connection. *Educational Leadership*, 51(8), 16-21.

Harding, Sandra (1986). *The science question in feminism.* Ithaca, NY: Cornell University Press.

Hein, George (1994). *Active assessment for active science.* Portsmouth, NH: Heinemann Press.

Hofstadter, Douglas R. (1985). *Metamagical themes: Questing for the essence of mind and pattern.* New York: Basic Books.

Hubbard, Ruth (1976). Rosalind Franklin and DNA. *Signs*, 2, 229-237

Hutchins, Pat (1986). *The door bell rang.* New York: Scholastic Inc.

Irigaray, Luce (1989). Is the subject of science sexed? In Nancy Tuana (Ed.), *Feminism and Science* (pp. 58-68). Bloomington, IN: Indiana University Press.

Irons, Calvin (1992). *Crocodile's oat.* San Francisco: Mimosa Publications.

Jacobs, Judith (1992). Women's learning styles and the teaching of mathematics. In Whitney Ransom & Meg Moulton (Eds.). *Math and science for girls* (pp. 42-55). Concord, MA: National Coalition of Girls' Schools.

Kahle, Jane Butler & Matyas, Marsha Lakes (1987). Equitable science and mathematics education: A discrepancy model. *Women: Their underrepresentation and career differentials in science and engineering.* Proceedings of a workshop. Washington DC: National Academy Press.

Keller, Evelyn Fox (1985). A world of difference. *Gender and science* (pp. 158-176). New Haven, CT: Yale University Press.

—— (1987). *A feeling for the organism: The life and work of Barbara McClintock.* San Francisco: W.H. Freeman.

—— (1990). Long live the differences between men and women scientists. *The Scientist,* 4(20), 15.

Kloosterman, Peter (1990). Attributions, performance after failure, and motivation in mathematics. In Elizabeth Fennema & Gilah C. Leder (Eds.), *Mathematics and gender* (pp. 96-127). New York: Teachers College Press.

Koch, Janice (1990). The science autobiography. *Science and Children,* 28(3), 42-43.

Koch, Janice & Blunck, Susan (1996). Breaking the mold: Celebrating individual difference in the science classroom. In Robert Yager (Ed.), *Science/technology/society as reform in science education* (pp. 119-130). Albany, NY: SUNY Press.

Lewis, Ellen (1993). Hers: Making a difference. *New York Times Magazine,* December 12.

Linn, Marcia (1992). Gender differences in educational achievement. *Sex Equity in Educational Opportunity, Achievement and Testing.* Princeton ,NJ: Educational Testing Service.

Lobel, Arnold (1970). *Frog and toad are friends.* New York: Scholastic Inc.

Macdonald, Anne L. (1992) *Feminine ingenuity: How women inventors changed America.* New York: Ballantine Books.

Malcom, Shirley & Kulm, Gerald (1991). *Science assessment in service of reform.* Washington, DC: American Association for the Advancement of Science.

Mathematical Sciences Education Board, National Research Council (1993). *Measuring what counts: A conceptual guide for mathematics assessment.* Washington, DC: National Academy Press.

Matyas, Marsha Lakes (1985). Obstacles and constraints on women in science: Preparation and participation in the scientific community. In Jane Butler Kahle (Ed.), *Women in Science: A report from the field* (pp. 77-102). Philadelphia: Falmer Press.

McIntosh, Peggy (1983). Interactive phases of curricular re-vision: A feminist perspective. Working Paper no. 124. Wellesley, MA: Wellesley Center for Research on Women.

Meyer, Margaret R. & Koehler, Mary S. (1990). Internal influences on gender differences in mathematics. In Elizabeth Fennema & Gilah C. Leder (Eds.), *Mathematics and gender* (pp. 60-94). New York: Teachers College Press.

National Coalition of Girls' Schools (1992). *Math and science for girls.* Concord, MA: National Coalition of Girls' Schools.

National Research Council (1995). *The national science education standards.* Washington, DC: National Academy Press.

Nelkin, Dorothy (1987). *Selling science: How the press covers science and technology.* New York: W.H. Freeman and Company.

Nichols, Rosalie S. & Kurtz, V. Ray (1994). Gender and mathematics contests. *Arithmetic Teacher,* 41(5), 238-239.

Quindlen, Anna (1993). Birthday girl: Teaching daughters to resist. *New York Times,* November 21, E17.

—— (1994). Barbie at 35. *New York Times,* September 10, 1994.

Raffalli, Mary (1994). Why so few women physicists? *New York Times,* January 9, ED26.

Riesz, Elizabeth D., McNabb, Terry F., Stephen, Sandra L., & Ziomek, Robert L. (1994). Gender differences in high school students' attitudes toward science: Research and intervention. *Journal of Women and Minorities in Science and Engineering,* 1, 273-289.

Rosser, Sue V. (Ed.) (1990). *Female-friendly science: Applying women's studies methods and theories to attract students.* New York: Pergamon Press.

—— (1990). What I learned from the bag lady scientist and the Nobel laureate James Watson. In Sue Rosser (Ed.). *Female-friendly science: Applying women's studies methods and theories to attract students* (pp. 106-111). New York: Pergamon Press

—— (1995). *Teaching the majority: Breaking the gender barrier in science, mathematics and engineering.* New York: Teachers College Press.

Rossiter, Margaret (1985). *Women scientists in America: Struggles and strategies to 1940.* Baltimore: Johns Hopkins University Press.

Sanders, Jo (1994). *Lifting the barriers: 600 strategies that really work to increase girls' participation in science, mathematics and computers.* Seattle: Jo Sanders Publications.

Sandler, Bernice (1991). Warming up the chilly climate. *Math and science for girls* (pp. 26-41). Concord, MA: National Coalition of Girls' Schools.

——, Silverberg, Lisa A., & Hall, Roberta M. (1996). *The chilly classroom climate: A guide to improve the education of women.* Washington, DC: National Association for Women in Education.

Sayre, Anne (1975). *Rosalind Franklin and DNA.* New York: Norton & Co.

Schwartz, Marilyn, and the Association of American University Presses (1995). *Guidelines for bias-free writing.* Bloomington, IN: Indiana University Press.

Shade, Barbara J. & New, Clara A. (1993). Cultural influences on learning: Teaching implications. In James A. Banks & Cherry A. McGee Banks (Eds.), *Multicultural education: Issues and perspectives* (pp. 317-331). Boston: Allyn and Bacon.

Sheehan, Kathryn & Waidner, Mary (1991). *The earth child.* Tulsa, OK: Council Oak Books.

Slobodkina, Esphyr (1989). *Caps for sale.* New York: Scholastic Inc.

Spertus, Ellen (1991). *Why are there so few female computer scientists?* Cambridge, MA: MIT Artificial Intelligence Laboratory, Technical Report 1315.

Steinberg, Adria (1994). When bright kids get bad grades. *The Best of the Harvard Education Letter*. Cambridge, MA: Harvard Graduate School of Education.

Tobias, Sheila (1993). *Overcoming math anxiety*. New York: W.W. Norton and Company.

Wiggins, Grant (1993). Assessment, authenticity, context and validity. *Phi Delta Kappan*, 75(3), 200-214.

ADDITIONAL MATHEMATICS AND GENDER RESOURCES

Burns, Marilyn (1977). *The good times math event book.* Worth, IL: Creative Publications.

—— (1987). *A Collection of math lessons.* San Francisco: The Math Solution Publications.

Downie, Diane, Slesnick, Twila, and Stenmark, Jean Kerr (1981). *Math for girls and other problem solvers.* Berkeley, CA: EQUALS, Lawrence Hall of Science.

Franklin, Margaret (1990). *Add-ventures for girls: Building math confidence.* Volume 1: Elementary teacher's guide. Volume 2: Junior high teacher's guide. Newton, MA: Women's Educational Equity Act Publishing Center, Education Development Center.

Mathematic Association of America (1996). *She does math.* Washington, DC: Mathematic Association of America.

Mitchell, M. (1978). *History of mathematics.* Reston, VA: National Council of Teachers of Mathematics.

Perl, Terri (undated). *Math equals: Biographies of women mathematicians and related activities.* Menlo Park, CA: Addison Wesley.

Reimer, W. & Reimer, L. (1990). *Mathematicians are people, too.* Palo Alto, CA: Dale Seymore Publications.

—— (1992). *Historical connections in mathematics.* Fresno, CA: AIMS Educational Foundation.

Stenmark, Jean Kerr, Thompson, Virginia, & Cossey, Ruth (1986). *Family math.* Berkeley, CA: Lawrence Hall of Science, University of Virginia.

Tobias, Sheila (1993). *Overcoming math anxiety* (2nd ed.). New York: W.W. Norton & Co.

Woodrow Wilson National Fellowship Foundation (1993). *Woodrow Wilson gender equity in mathematics and science congress.* Princeton, NJ: Woodrow Wilson National Fellowship Foundation.

Zavasky, Claudia (1994). *Fear of math: How to get over it and get on with your life.* New Brunswick, NJ: Rutgers University Press.

ADDITIONAL SCIENCE AND GENDER RESOURCES

Abir-Am, Pnina and Outram, Dorinda (1987). *Uneasy careers and intimate lives: Women in science, 1789-1979.* New Brunswick, NJ: Rutgers University Press.

Alic, Margaret (1986). *Hypatia's heritage: A history of women in science from antiquity through the nineteenth century.* Boston: Beacon Press.

American Association for the Advancement of Science (1990). *Science for all Americans.* New York: Oxford University Press.

—— (1993). *Benchmarks for science literacy: Project 2061.* New York: Oxford University Press.

Blier, Ruth (1984). *Science and gender: A critique of biology and its theories on women.* New York: Pergamon Press.

—— (1986). *Feminist approaches to science.* New York: Pergamon Press.

Byrne, Eileen (1993). *Women and science: The snark syndrome.* Bristol, PA: Falmer Press / Taylor and Francis.

Hubbard, Ruth (1990). *The politics of women's biology.* New Brunswick, NJ: Rutgers University Press.

Keller, Evelyn Fox (1983). *A feeling for the organism: The life and work of Barbara McClintock.* New York: W.H. Freeman and Company.

—— (1985). *Reflections on gender and science.* New Haven, CT: Yale University Press.

—— (1990). Long live the differences between men and women scientists. In *The Scientist,* October 15, 1990.

McIlwee, Judith & Robinson, J. Gregg (1992). *Women in engineering: Gender, power, and workplace culture.* Albany, NY: SUNY Press.

National Coalition for Girls' Schools (1993). Girls in the physical sciences. Concord, MA: National Coalition for Girls' Schools.

National Research Council (1995). *The national science education standards.* Washington, DC: National Academy Press.

Rosser, Sue V. (1986). *Teaching science and health from a feminist perspective.* New York: Pergamon Press.

Rosser, Sue V. & Kelly, Bonnie (1994). *Educating women for success in science and mathematics.* Columbia, SC: Division of Women's Studies, University of South Carolina.

Sime, Ruth Lewin (1996). *Lise Meitner: A life in physics.* Berkeley: University of California Press.

Tuana, Nancy (Ed.). (1989). *Feminism and science.* Indianapolis: Indiana University Press.

Quinn, Susan (1995). *Marie Curie: A life.* New York: Simon and Schuster.

AUDIOVISUAL RESOURCES

Posters of Women in MST

CIBA-GEIGY Public Affairs
444 Sawmill River Road
Ardsley, NY 10502
(914) 479-5000; fax 785-4524

Consortium for Educational Equity
4090 Livingston Campus
Rutgers University
New Brunswick NJ 08903
(908) 445-2071; fax -0027

National Women's History Project
7738 Bell Road
Windsor CA 95492
(707) 838-6000; fax -0478

Organization for Equal Education of the Sexes
PO Box 438
Blue Hill ME
(207) 374-2489; fax -2890

Women's Educational Equity Act
(WEEA) Publishing Center
Education Development Center
55 Chapel Street
Newton MA 02158
(800) 225-3088; fax (617) 332-4318

Videotapes

Association for Computing (1995). "Minerva's Machine: Women and Computing." New York: Association for Computing Machinery, about an hour.

American Association for the Advancement of Science (1991). "Girls, Science and Mathematics." Washington, DC: AAAS, 17 minutes.

American Association of University Women (1994). "Girls Can." Washington DC: AAUW, 16.5 minutes.

Intelecom (undated). "Breaking Through: Women in Science." Pasadena CA: Intelecom, 29 minutes.

National Coalition of Girls' Schools (1995). "Girls and Technology: Expect the Best." Concord MA: National Coalition of Girls schools, 22 minutes. Accompanied by resource guide.

Northeastern University Women's Programs, College of Engineering (1987). "Women in Engineering. Boston: Northeastern University, length not given but fairly short.

Western Illinois University (undated). "Connecting the Past with the Future: Women in Mathematics and Science." 45-minute videos on Ada Lovelace, Mary Sommerville, Maria Gephard Meyer, and one other. Macomb, IL: Western Illinois University.

See also: Bibliographies

BIBLIOGRAPHIES

Hulme, Marylin (1995). *Audiovisual materials: Films, videotapes, slides & audiotapes.* fifth edition. New Brunswick, NJ: Consortium for Educational Equity, Rutgers University.

—— (1995). *Equity materials in mathematics, science and technology: A resource guide,* third edition. Philadelphia: Research for Better Schools.

Sanders, Jo and Rocco, Starla (1994). *Bibliography on gender equity in mathematics, science and technology: Resources for classroom teachers.* Seattle: Jo Sanders Publications.

Wellesley Center for Research on Women (1992). *A bibliography of research on girls in U.S. public schools, kindergarten through grade 12.* Wellesley, MA: Wellesley Center for Research on Women.

ELECTRONIC RESOURCES (Current as of 1996)

Biographies of female astronauts. This one is for Eileen Collins; others are available for Sally Ride, Mae Jemison, Judith Resnick, and others.
http://www.jsc.nasa.gov/bios/htmlbios/collins.html

Directory of organizations encouraging women in science and engineering
http:// www.nas.edu/cwse

Diversity in education
http://www.inform.umd.edu:8080/ed/res/topic/diversity

Educational equity listserve, "EDQUITY." Send SUBSCRIBE EDEQUITY to:
majordomo@confer.edc.org

Ensuring Equity and Excellence in Mathematics
http://www.ncrel.org.ncrel/sdrs/areas/issues/content/cntareas/math/ma100.html

Ensuring Equity and Excellence in Science
http://www.ncrel.org/ncrel.sdrs/areas/issues/content/cntareas/science/sc200.html

Feminist Majority Foundation
http://www.feminist.org

Forums, women's and gender issues and lists
http://www.unix.umbc.edu/-korenman/wmst/forums

Gender equity in education listserve, "GENED." Send SUBSCRIBE GENED to:
majordomo@acpub.duke.edu

Gender issues directory
http://cpsr.org/dox/gender.html

Gender issues in sport
http://www.arcade.uiowa.edu/proj/ge/

MIT World Wide Web page on women in science, mathematics, and technology
http://www.ai.mit.edu/people/ellens/gender.html

Organization for the Equal Organization of the Sexes
http://media1.hypernet.com/oees.html

Pleiades Networks, a World Wide Web site for women: An illustrated guide to the Internet, directories of women's organizations and resources, calendar of events, discussion groups
http://www.pleiades-net.com/

Regional Alliance, which covers equity topics
http://www.hub.terc.edu

Teacher Talk for preservice secondary education students, also distributed in print
http://education.indiana.edu/cas/tt/tthmpg.html

Women's studies listserve. Send SUBSCRIBE WMST-L to:
listserv@umdd.umd.edu

Women in Technology web site
http:// gseweb.harvard.edu/TIEWeb/STUDENTS/STUDENTGROUPS/WIT/wit.html

ORGANIZATIONS FOR WOMEN IN MST

American Association of University Women
1111 16th Street, NW,
Washington, DC 20036
(202) 785-7700; Fax 872-1425
<info@mail.aauw.org>
http://www.aauw.org

Association for Women in Mathematics (AWM)
4114 Computer & Space Sciences Building
University of Maryland
College Park, MD 20742-2461
(301) 405-7892
<awm@math.umd.edu>
http://www.math.neu.edu/awm

Association for Women in Science (AWIS)
1200 New York Avenue, N.W., Suite 650
Washington, DC 20005
202/326-8940; fax -8950
<awis@awis.org>

Consortium for Educational Equity
Equity Assistance Center
Rutgers, The State University of New Jersey
4090 Livingston Campus
New Brunswick, NJ 08903
(908) 445-2071; fax -0027

EQUALS / Family Math
Lawrence Hall of Science
University of California
Berkeley, CA 94720
(510) 642-1823; fax 643-5757
<equals@uclink.berkeley.edu>

Math/Science Network - Expanding Your Horizons conferences for teenage girls
Mills College
5000 MacArthur Blvd.
Oakland CA 94613
(510) 430-2222; fax -2090
<msneyh@mills.edu>

National Coalition of Girls' Schools
228 Main Street
Concord, MA 01742
(508) 287-4485; fax -6014
<ncgs@ ncgs.org>
http://www.tiac.net/users/ncgs

National Women's History Project
7738 Bell Road
Windsor, CA 95492
(707) 838-6000; fax -0478
<nwhp@ aol.com>
http:// www.nwhp.org

Organization for Equal Education of the Sexes
P.O. Box 438
Blue Hill, ME 04614
(207) 374-2489

Society of Women Engineers
120 Wall Street, 11th floor
New York, NY 10005-3902
(212) 509-9577; fax -0224
<71764.743@compuserve.com>

Wellesley Center for Research on Women
Wellesley College
Wellesley, MA 02181
(617) 283-2500; fax -2504

Women's Educational Equity Act (WEEA) Publishing Center
Education Development Center, Inc.
55 Chapel Street, Suite 275
Newton, MA 02160
(800) 225-3088; fax (617) 332-4318
<weea@edc.org.
http://www.edc.org.ceec.weea

Women in Mathematics Education
Can be reached via Charlene and James Morrow at the SummerMath program, Mt. Holyoke College; see p. 112

Women in Science Survey (WISS)
Contact: Dr. Lynn Mulkey
Sociology Department
Hofstra University
Hempstead, NY 11550
(516) 463-5000

PROGRAMS FOR GIRLS OR WOMEN IN MST

Brookhaven Women in Science
Brookhaven National Lab
Upton, NY11973
(516) 282-2123

The program has a speakers' bureau in which professional women visit schools and organizations to encourage girls to consider careers in science, mathematics and engineering, an annual scholarship given to encourage women to resume their formal studies in the sciences, engineering, or mathematics, and a prize to recognize substantial promise and achievement by a woman graduate student in physics at the State University of New York at Stonybrook.

Center for Family Involvement in Schools
Consortium for Educational Equity
Rutgers, The State University of New Jersey
Building 4090, Livingston Campus
New Brunswick, NJ 08903
(908) 445-2071; fax -0027

This program provides schools with equitable model programs and training to involve parents actively in the intellectual and academic achievement of their children. It especially offers Family Math, Family Science, and Family Tools and Technology.

EQUALS / Family Math
Lawrence Hall of Science
University of California
Berkeley, CA 94720
(510) 642-1823; fax 643-5757
<equals@uclink.berkeley.edu>

These are workshops in English and Spanish for parents and children emphasizing mathematics and equity and featuring math activities parents and children can do together. There are also workshops for mathematics teachers (EQUALS in Mathematics Workshops, SEQUALS, and Investigations). A guidebook is available in English and Spanish.

EUREKA Teen Achievement Program
College of Education
Brookyn College
2900 Bedford Avenue, Room 211
Brooklyn, NY 11210
(718) 951-5214

This is a summer career and leadership development program to encourage precollege minority girls in grades 8 through 12, primarily from economically disadvantaged families, to pursue high achievement in mathematics, science, computers, and sports, and eventually to consider careers in related fields. This program emphasizes the connections between formal and informal education.

Expanding Your Horizons
Math/Science Network
Mills College
Oakland CA 94613
(510) 430-2222; fax -2090
<msneyh@mills.edu>

This is a one-day hands-on Saturday conference for girls in grades 6 through12 with women role models leading the workshops. There are also sessions for teachers and parents. Conferences are organized by grassroots groups. A conference guide is available. This is a popular, frequently implemented, and relatively easy-to-follow model.

Females Achieving Mathematics Equity (FAME)
Fairfax County Public Schools
Burholder Administrative Center
10700 Page Avenue
Fairfax, VA 22030
(703) 246-2502

This is a one-week summer program for minority middle-school girls in which students visit local businesses and participate in hands-on mathematics activities. The goal of the program is to encourage increased enthusiasm for higher-level mathematics courses.

Gender/Ethnic Expectations and Student Achievement (GESA)
GrayMill Foundation
22821 Cove View Street
Canyon Lake, CA 92587
(909) 244-5165

This is a popular on-site training program for teachers that addresses gender and racial bias throughout the curriculum and in all grades, delivered by experienced trainers.

Girls Count
225 East 16th Avenue, Suite 475
Denver, CO 80203
(303) 832-6600; fax -7331
<girlscount@aol.com>

Girls Count helps the adults in girls' lives to support girls' academic achievement and career aspirations especially in mathematics, science, and technology through training programs, public education, and public policy initiatives. Training programs are available to communities nationwide.

MESA: Mathematics, Engineering, Science Achievement
312 McLaughlin Hall
College of Engineering
University of California at Berkeley
Berkeley, CA 94720-1702
(510) 642-2041; fax 643-5600

This is a program that prepares students in grades five through twelve, primarily minority, for academic and professional leadership positions through tutoring, field trips to industry, guest speakers, workshops, and college tours. Classes are held on campuses on Saturdays.

National SEED Project: Seeking Educational Equity and Diversity
Wellesley College Center for Research on Women
Wellesley, MA 02181
(617) 283-2522; fax -2504

A week-long SEED summer Leaders' Workshop prepares teachers to hold year-long reading groups with other teachers to make school curricula more gender-fair and multicultural. It has been in operation for about 10 years.

Northwest EQUALS / Family Science
P.O. Box 1491
Portland State University
Portland, OR 97207
(800) 547-8887

Family Science is a K-8 national outreach program to teach science by having children and parents learn and enjoy science together. It addresses the underrepresentation of women and ethnic and racial minorities in the sciences. A 10-hour inservice program and a book of activities are available.

Operation SMART: Science, Math, and Relevant Technology
Girls Incorporated
441 West Michigan Street
Indianapolis, IN 46202-3233
(317) 634-7546; fax -3024
<hn3580@handsnet.org>

This is an informal education program started in 1985. Local clubs use hands-on, participatory activities to stimulate girls' interest in science, math, and technology. There is strong emphasis on involving children who have traditionally been left out of math and science: ethnic and racial minorities and children with disabilities.

SummerMath
Drs. James and Charlene Morrow
Mount Holyoke College
South Hadley, MA 01075-1441
(413) 538-2608; fax -2002

SummerMath is geared toward college-bound girls in grades eight through twelve of average math ability. The six-week summer program focuses on improving problem-solving skills and building confidence. Several undergraduates also serve as Residential and Teaching Assistants in the program. Applications for RA/TA's are due in February; those for girls are due in late May.

ASSESSMENT RESOURCES

Materials and Associations

A sampler of mathematics assessment
Bureau of Publications
California Department of Education
PO Box 271
Sacramento, CA 95812-0271
(916) 445-1260

Assessment alternatives in mathematics by Jean Stenmark (1989)
Attn: Assessment Booklet
EQUALS
Lawrence Hall of Science, Room 240
University of California
Berkeley, CA 94720
(510)·642-1823

Center for the Study of Testing, Evaluation, and Educational Policy
Boston College
Campion Hall, Room 323
Chestnut Hill, MA 02167
(617) 552-4521

Educational Testing Service
Rosedale Road
Princeton, New Jersey 08541
(609) 921-9000

Fair Test
342 Broadway
Cambridge, MA 02139
(617) 864-4810

National Assessment of Educational Progress (NAEP)
US Department of Education
555 New Jersey Ave. NW
Washington, DC 20208-5653
(202) 219-1761

Mathematics assessment myths, models, good questions, and practical suggestions (1991), Jean Stenmark (ed.)
National Council of Teachers of Mathematics
1906 Association Drive
Reston, VA 22091-1593
(703) 620-9840

National Science Teachers Association
1840 Wilson Boulevard
Arlington, VA 22201
(703) 243-7100

New Standards Project
Learning Research and Development Center
University of Pittsburg
3939 O'Hara Street
Pittsburg, PA 15260
(412) 624-8319

Testing Companies and Publishers

Continental Press
520 East Bainbridge
Elizabethtown, PA 17022
(800) 233-0759

Curriculum Associates
5 Esquire Road
North Billerica, MA 01862
(800) 225-0248

CTB Mc Graw-Hill
20 Ryan Ranch Road
Monterey, CA 93940
(800) 538-9547

Psychological Corporation
555 Academic Court
San Antonio, TX 78204
(800) 228-0752

Riverside Publishing
8420 West Bryn Mawr Avenue
Chicago, IL 60631
(800) 323-9540

Steck-Vaughn
P.O. Box 26015
Austin, TX 78755
(800) 531-5015

Education Departments

California Department of Education
721 Capitol Mall Box 944272
Sacramento, CA 94244 -2720
(916) 657-2451

Maryland State Department of Education
200 W. Baltimore Street
Baltimore, Maryland 21201-2595
(410) 767-0100

Vermont Institute of Mathematics, Science, and Technology
Box 310
Waterbury Center, VT 05677
(802) 244-8788

Kentucky Department of Education
500 Mero Street
Frankfort, KY 40601
(502) 564-4770

University of the State of New York
State Education Department
Albany, NY 12234
(518) 474-3900